Circuit Analysis
Laboratory Workbook

Synthesis Lectures on Electrical Engineering

Editor
Richard C. Dorf, *University of California, Davis*

Circuit Analysis Laboratory Workbook

Teri L. Piatt and Kyle E. Laferty

ISBN: 978-3-031-00890-0   paperback
ISBN: 978-3-031-02018-6   ebook

DOI 10.1007/978-3-031-02018-6

A Publication in the Springer series
SYNTHESIS LECTURES ON ELECTRICAL ENGINEERING

Lecture #4
Series Editor: Richard C. Dorf, *University of California, Davis*
Series ISSN
Print 1559-811X   Electronic 1559-8128

Circuit Analysis
Laboratory Workbook

Teri L. Piatt and Kyle E. Laferty
Wright State Univesity

SYNTHESIS LECTURES ON ELECTRICAL ENGINEERING #4

ABSTRACT

This workbook integrates theory with the concept of engineering design and teaches troubleshooting and analytical problem-solving skills. It is intended to either accompany or follow a first circuits course, and it assumes no previous experience with breadboarding or other lab equipment. This workbook uses only those components that are traditionally covered in a first circuits course (e.g., voltage sources, resistors, potentiometers, capacitors, and op amps) and gives students clear design goals, requirements, and constraints. Because we are using only components students have already learned how to analyze, they are able to tackle the design exercises, first working through the theory and math, then drawing and simulating their designs, and finally building and testing their designs on a breadboard.

KEYWORDS

circuits, circuit design, analog circuit lab, electronics, electronic circuits, electrical engineering, circuits labs, operational amplifiers (op amps), analog circuits, circuit analysis lab, circuit analysis

Contents

Preface

Philosophy This workbook integrates the theory covered in a first circuits course with the concept of engineering design and teaches troubleshooting and analytical problem-solving skills. Traditionally, circuits labs give step-by-step instructions for building a circuit and taking measurements from a given schematic to demonstrate physically the theory learned in class, point out differences between measured values and theoretical calculations, and teach measurement techniques.

In contrast, this workbook asks students to use the theory they are learning in class to design circuits that are simple and functional. In doing so, they learn design techniques, develop their analytical problem-solving skills, and gain a better understanding of theoretical principles. So, instead of demonstrating theory, these labs ask students to apply theoretical principles to a design problem.

Approach This workbook uses only those components that are traditionally covered in a first circuits course and gives students clear design goals, requirements, and constraints. It begins with very simple design exercises and progresses to fairly complex circuits, all while using only basic components, such as resistors, potentiometers, capacitors, and operational amplifiers. Because we are using only components students have already learned how to analyze, they are able to complete the design exercises, first working through the theory and math, then drawing and simulating their designs, and finally building and testing their designs in the lab.

Since they are not merely wiring a given circuit, they must understand the theoretical concepts in order to complete their designs. The process of troubleshooting their circuits then teaches them both analytical problem-solving skills and the difference between theory (circuit works on paper) and real life (wired circuit is not accurate).

Audience This workbook is meant to accompany a first course in linear, analog circuit analysis that covers topics such as Kirchhoff's laws, Ohm's law, node analysis, equivalent circuits (e.g., Thevenin), superposition, first- and second-order circuits, and AC steady-state analysis. Components covered include resistors, capacitors, inductors, operational amplifiers, and voltage sources (DC and AC).

Background We are assuming students have no previous experience using breadboards, multimeters, or other lab equipment.

Student Workload We are assuming that students will spend 1-2 hours on each pre-lab and will have the pre-lab completed before their scheduled lab period. If students are prepared, the in-lab portions and post-lab questions can be completed in a 2-hour lab period.

TA Workload We have found it helpful for the TAs to spend 10-15 minutes at the beginning of the lab period giving a short lecture or demonstration to the class. This should include a demonstration on how to use a new piece of equipment, safety reminders, reviews of measurement techniques (e.g., measuring voltage versus measuring current), and a list of troubleshooting ideas (we have provided some in Appendix C).

Some of the in-lab exercises begin with a statement like, "Your TA will demonstrate how to use the function generator." This is another way we have deviated from a traditional circuits lab: we do not have pictures of equipment with detailed wiring diagrams. Students watch their TA give a demonstration, see the diagrams the TA draws on the board, and then they are ready to tackle that week's circuit build.

In addition, the TAs are also expected to supervise students' work throughout the lab period, and each lab has at least one "TA Verification" space for the TA to sign off on students' work.

Organization Since these lab exercises are constrained to use only those components students have already learned how to analyze in class, this workbook follows the traditional outline of most circuits textbooks. This means that the first six labs use only voltage supplies and resistive elements (e.g., resistors, potentiometers, and thermistors). We use these early labs to introduce the basics, such as using a breadboard and digital multimeter, identifying resistor values, and wiring resistors in parallel and series. These early labs also introduce the idea of design work, setting out problems that get gradually more difficult and include design requirements and constraints.

The schedule in more detail is as follows.

- Lab 1 has no pre-lab and is meant to be completed the first week of the term.

- Lab 2 is dedicated to students learning the basics of drawing and simulating a circuit using Multisim. This lab is easily skipped, if there is not time or if Multisim skills are not a part of your curriculum. If included, Appendix B gives instructions for basic Multisim work. Also, we usually allow two weeks for this lab, as the pre-lab is time-consuming for those students not already familiar with Multisim.

- Labs 3, 4, and 5 use only resistors, potentiometers, DC voltage supplies, and multimeters. These labs also introduce the concept of design goals and constraints.

- Lab 6 introduces the idea of a resistive sensor (in this case, a thermistor) used in a Wheatstone bridge.

- Lab 7 introduces the use of ICs and operational amplifiers, as many circuit classes do not begin discussing op amps until five or six weeks into the term. All of the remaining labs use op amps in a series of design challenges.

- Lab 8 brings back the Wheatstone bridge from Lab 6, with the new challenge of adding an amplifier and indicator light to improve the circuit's functionality.

- Labs 9 and 10 challenge students to design and build an analog calculator.

- Labs 11 and 12 introduce AC inputs, function generators, and oscilloscopes and ask students to integrate phase shift into their designs (Lab 11) and to design two simple, first-order filters (Lab 12).

Required Lab Equipment Our labs rely on standard circuits lab equipment. Each lab station should have:

- a breadboard,

- a power supply,

- a digital multimeter (for measuring voltage and current),

- a function generator (for producing a sinusoidal and triangle-wave supply voltages), and

- an oscilloscope.

Each lab station should have access to:

- standard $\frac{1}{4}$ W or $\frac{1}{2}$ W resistors, with values from 100 Ω - 1 MΩ,

- at least five 10 kΩ potentiometers,

- operational amplifiers (LM741),

- red and/or green LEDs ,

- capacitors: 10 μF, 0.33 μF, 0.03 μF, 0.02 μF, 0.0075 μF, 0.002 μF,

- two audio jacks (we use Kycon STX-3120 3.5 mm PCB mount),

- ear buds or headphones (students can bring their own), and

- one NTC thermistor (10 kΩ at 25°C; we use Dale/Vishay 01C1002KP, whose data sheet is provided in Appendix F).

In addition, the TAs should have access to a heat gun and a lamp (incandescent bulb). These are used to heat up the thermistors, and alternative means, such as a hair dryer, blowing on the thermistors, or holding them between two fingers, would also be fine.

Teri L. Piatt and Kyle E. Laferty
June 2017

Acknowledgments

The authors would like to thank the students who volunteered to test early versions of all of these labs: Thomas Beard, Jeremy Hong, Nick Miller, Zach Quach, Christian Stewart, and Emma Sum. We are also very grateful to our proofreader and sounding-board, Dr. Greg Reich, and the great team at Morgan & Claypool Publishers.

Teri L. Piatt and Kyle E. Laferty
June 2017

LAB 1

This is my Breadboard

Purpose: In this lab, you will learn to use basic electronic lab equipment, including a breadboard, a power supply, and a digital multimeter.

In the lab: The main equipment used in this lab will be breadboards, power supplies, multimeters, and resistors. *Breadboards* are a prototyping tool used to hold and connect circuit components without solder (see Figure 1.1). Figure 1.2 shows graphically how the breadboard's internal connections are laid out. The top two and bottom two rows of the breadboard shown in Figures 1.1 and 1.2 are connected along the length of the breadboard and are referred to as power rails. Power rails are a way to provide the same supply voltage to multiple, different parts of a circuit and are often labeled with a "+" and "−" symbol at the end of each row. We will be using the power rails for more complex circuits later in the semester.

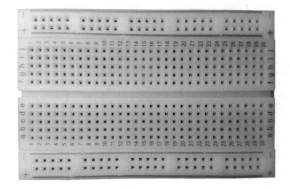

Figure 1.1: A small breadboard [1].   Figure 1.2: A breadboard's internals.

The main part of the breadboard is the middle area, which has rows labeled a–j in Figures 1.1 and 1.2. Both the bottom and top set of five holes in a column are seperately interconnected, and none of the horizontal holes in a row are connected to each other. So, for example, if you place one lead of a component in row "b" and the other lead in row "e" of the same column, then you have effectively wired those two leads to each other, shorting that component. Generally, it is a good idea to build your circuit with the components situated parallel to the power rails.

There are two main types of power sources used in the lab, DC and AC. DC stands for direct current and provides a steady voltage at one value. AC stands for alternating current. The most common type of AC is a sinusoidal signal, but square waves and triangle waves are other examples of AC. Because AC signals can take many forms, the source is often referred to as a function generator. Some schematic symbols for voltage sources are shown in Figure 1.3. Figure 1.4 shows three symbols for another important breadboard connection, *ground*, which is a reference node necessary for a circuit to work properly and where the voltage is always zero.

Figure 1.3: Different symbols for voltage sources. The left two are both DC source schematic symbols. The right symbol is an AC source schematic symbol.

Figure 1.4: Different symbols for ground.

Resistors are the most basic circuit component, and their schematic symbol is shown in Figure 1.5. They *resist* current, and without them a lot of circuits would not work. Different resistors have different resistance values, measured in Ohms (Ω). Engineers use a color code to quickly decipher the value of a resistor (see Table 1.1). A resistor has four color bands encircling it. The first two bands indicate the value of the resistor, and the third color band indicates its order of magnitude. So, if a resistor has color bands of violet, yellow, and red, the resistance value is $74 * 10^2$. This value would be written as 7.4 kΩ. (It is not proper to leave the 10^2 in the answer). The fourth color band indicates the resistor's tolerance. The most common tolerance bands are gold and silver; gold indicates a $\pm 5\%$ tolerance and silver indicates a $\pm 10\%$ tolerance.

Figure 1.5: Resistor schematic symbol.

Table 1.1: Resistor color code

Resistor Color Code									
Black	Brown	Red	Orange	Yellow	Green	Blue	Violet	Grey	White
0	1	2	3	4	5	6	7	8	9

Given the following color bands, find the resistance value:

- Red, Red, Red: _____

- Brown, Black, Brown: _____

- Blue, Orange, Green: _____

- Green, Brown, Orange: _____

- Brown, Green, Violet: _____

- White, Brown, Green: _____

- Gray, Red, Black: _____

- Yellow, Violet, Red: _____

- Red, Yellow, Black: _____

Figure 1.6: Comic relief [2].

Build the circuit shown in Figure 1.7 on your breadboard.

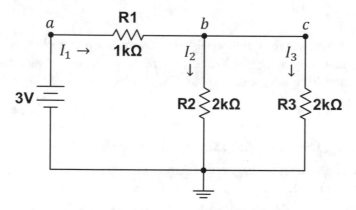

Figure 1.7: Lab 1 circuit.

Use the digital multimeter to measure voltage and current. To measure voltage, set the dial of the multimeter to measure DC voltage. Then put the multimeter's red lead on one terminal of the resistor (or component) and the multimeter's black lead on the other terminal of the resistor (or component). For example, if you want to measure the voltage across R_1, place the red lead at point a and the black lead at point b. This is V_{ab}, and the value displayed will be in volts. Now place the red lead at b and the black lead at a. This is V_{ba}.

To measure current, set the dial of the multimeter to measure mA. Break the circuit at the desired spot, and connect the leads of the multimeter across the open. That is, complete the circuit with the multimeter. The value displayed will be in mA. **NOTE: measuring current has to be done in a very specific way. If you try to measure current the way you measure voltage, you will blow a fuse in the multimeter, and YOU will be responsible for replacing it!** For more instructions on measuring voltage and current, please refer to Appendix A.

Fill in Table 1.2 with your measured values, and demonstrate at least one of your measurements to your TA.

Table 1.2: Data for circuit in Figure 1.7

I_1 (mA)	I_2 (mA)	I_3 (mA)	V_a (V)	V_b (V)	V_c (V)	V_{ab} (V)	V_{bc} (V)	V_{ba} (V)

TA Verification: _____

Post-lab questions: Answer the questions listed below on a separate piece of paper. Make sure that your handwriting is legible! When you are finished, staple everything together, and turn in this completed lab packet to your TA.

1. Use each resistor's tolerance to calculate the range of potential values for R_1, R_2, and R_3. For example if a resistor is 5.1 kΩ $\pm$10%, then that resistance could potentially be anything from 4590–5610 Ω. Find this range for each resistor in the circuit.

2. Use your current and voltage measurements to calculate the resistance of each resistor (use Ohm's Law).

3. Do these calculated values match each resistor's labeled resistance? If not, do your calculated values fall within the resistor's tolerance range?

LAB 2

Introduction to Multisim

Purpose: In this lab, you will learn how to draw and simulate basic circuits in Multisim.

Pre-lab: Multisim is a powerful circuit design and simulation program. It can help you test and refine your circuit design before you build a prototype. It can also help you understand what's happening in a circuit by allowing you to simulate the circuit and measure voltages, currents, and other values at any point. This reduces or eliminates the need for you to repeatedly solve a complex circuit by hand every time you make a change.

Using the information provided in Appendix B complete the following three drawings, measurements, and calculations.

1. For the circuit shown in Figure 2.1 below, calculate the current, I, in the loop and V_{ab}.

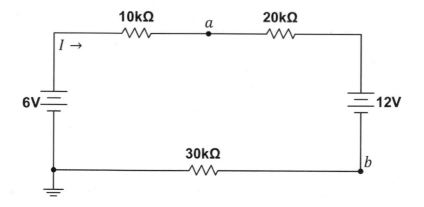

Figure 2.1: Lab 2 pre-lab circuit 1.

Enter your answers here, and show your calculations in the work area:

Calculated $I = $ _____   Calculated $V_{ab} = $ _____

Work area:

Use Multisim to draw the schematic shown in Figure 2.1. Simulate your circuit and use a current probe to measure the current in the loop and use the differential voltage probe to measure V_{ab}. Enter those measured values here, and print out your drawing to turn in with this lab.

Multisim I = _____   Multisim V_{ab} = _____

2. Use Multisim to draw the schematic shown in Figure 2.2. When you simulate the circuit, -12 V should appear on the voltmeter. Print out your drawing to turn in with this lab.

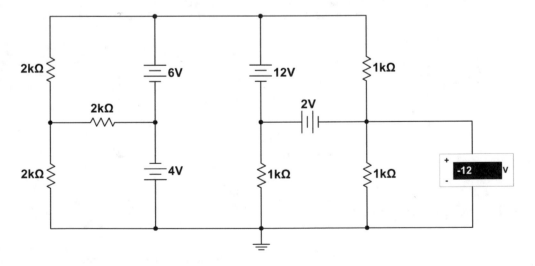

Figure 2.2: Lab 2 pre-lab circuit 2.

3. Use Multisim to draw the circuit shown in Figure 2.3. Then label Figure 2.3 with the direction of each branch current and the polarity for each resistor (you do not need to add these labels to your Multisim drawing). Simulate your circuit, and, using voltage and current probes, fill in the values for each component's voltage and current in Table 2.1.

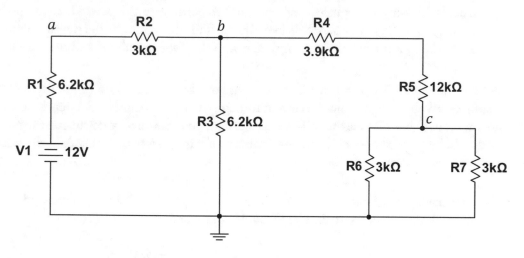

Figure 2.3: Lab 2 pre-lab circuit 3.

Table 2.1: Circuit 3 Multisim measured and calculated values

Component	Voltage (V) (Measured)	Current (mA) (Measured)	Power (mW) (Calculated)	Power (mW) (Measured)
R_1 =				
R_2 =				
R_3 =				
R_4 =				
R_5 =				
R_6 =				
R_7 =				
V_1 =				
		Total Power:		

Once that is complete, use these values to calculate the power *dissipated* by each component. (Pay attention to the direction of the current and remember the passive sign convention.)

Now add power probes to your Multisim drawing and simulate. Enter these measured power values in Table 2.1. Do the power values shown on the probes match those you calculated using the voltages and currents? According to the power law, the power dissipated = the power supplied in a circuit. Sum the values in the power columns. Does the power law hold true for the power values you calculated and measured?

In the lab: Build circuit 3 (shown in Figure 2.3) on your breadboard. Use the multimeter to measure the voltage across and the current through each component, and enter these measurements in Table 2.2. Use the voltages and currents you measured to calculated the power dissipated by each component, and enter these values in the table, as well. Add the values in the power column.

If you are having problems getting your circuit to work, see Appendix C for some ideas on how to troubleshoot your circuit. Demonstrate your working circuit to your TA.

TA Verification: _____

Table 2.2: Circuit 3 breadboard measured and calculated values

Component	Voltage (V) (Measured)	Current (mA) (Measured)	Power (mW) (Calculated)
$R_1 =$			
$R_2 =$			
$R_3 =$			
$R_4 =$			
$R_5 =$			
$R_6 =$			
$R_7 =$			
$V_1 =$			
		Total Power:	

Post-lab questions: Answer the questions listed below on a separate sheet of paper. Make sure that your handwriting is legible! When you are finished, staple everything together and turn in this completed lab packet to your TA.

1. Why didn't your power values add up to zero?

2. How close were your in-lab measured values (Table 2.2) to your Multisim measured values (Table 2.1)?

3. For the circuit that you built (circuit 3), use your measured data to verify Kirchhoff's current law at nodes a, b, and c. Show your work!

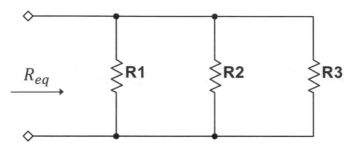

Figure 3.2: Three resistors connected in parallel.

1. Design a circuit with an equivalent resistance of 9 Ω using all three of the following resistor values: 4 Ω, 6 Ω, and 12 Ω. Draw your circuit in the work area.

   Work area:

2. Design a circuit with an equivalent resistance of 8 Ω using all four of the following resistor values: 2 Ω, 3 Ω, 4 Ω, and 12 Ω. Draw your circuit in the work area.

Work area:

3. Design a circuit with an equivalent resistance of 8 Ω using all six of the following resistor values: 3 Ω, 4 Ω, 4 Ω, 5 Ω, 6 Ω, and 42 Ω. Draw your circuit in the work area.

Work area:

Draw this third circuit in Multisim, and connect your network of resistors to a 1 V source. Simulate your circuit, and, using a current probe, measure the current leaving the source. Print your circuit with the current value visible. Using the source voltage, source current, and Ohm's law, calculate the equivalent R for this network of resistors. Does this value match the equivalent resistance you were trying to achieve? (Show your work in the work area.)

Work area:

In the lab: Determine the color bands for resistors having the following resistance ranges:

1. 2.97–3.63 MΩ

2. 6460–7140 Ω

3. 44.65–49.35 Ω

4. Using the following resistor values provided by your TA, design a circuit with an equivalent resistance of $R_{eq} = $ _____

   $R_1 = $ _____

   $R_2 = $ _____

   $R_3 = $ _____

   $R_4 = $ _____

   $R_5 = $ _____

Draw your circuit in the work area and have your TA verify that your design is correct.

Work area:

TA Verification: _____

5. If your network of resistors is connected to a 10 V source, calculate what current you would expect to see leaving the source. Enter your calculated current here, and show your work in the work area.

Calculated I_{in} = _____

Work area:

6. Assuming that your resistors all have a ±5% tolerance, calculate a maximum and minimum value for the current that you expect to see. Enter those values here, and show your calculations in the work area.

$$\text{_____} \leq I_{in} \leq \text{_____}$$

Work area:

7. Wire your circuit on your breadboard, and connect a 10 V source to your network of resistors. Measure the current leaving the source. Record the current, and demonstrate your measurement for your TA.

Measured I_{in} = _____   TA Verification: _____

8. On your breadboard, replace your network of resistors with a single resistor, as close to your R_{eq} as you can find. Measure the current leaving the source, record it here, and demonstrate your measurement for your TA.

Single-resistor I_{in} = _____   TA Verification: _____

9. Up to this point, you have been doing a very simple design problem. Being able to solve design problems is a critical skill for engineers. A design problem gives you an end goal, such as a minimum or maximum current or power to be delivered, and a set of constraints, such as limited tools, space, or money. So far in this lab, you have been given a design goal (e.g., the desired equivalent resistance) and constraints (e.g., the resistors you must use).

   Let's try a slightly different sort of design problem: design a circuit that will draw 2 mA of current from a 5 V source. So, your design goal is $I_{in} = 2$ mA, and your only constraint is that you must use a 5 V source. With no other constraints, this requires a single calculation! Draw your circuit in the work area.

   Work area:

Unfortunately, the resistor value you require is not a standard value. So, complete this design problem with this additional constraint: you may use as many resistors as necessary, but you may not use any resistors smaller than 1.5 kΩ. A list of standard resistor values can be found in Appendix D. Draw your circuit in the work area.

Work area:

Build this circuit and measure the current leaving the source. Record the current value here, and demonstrate your measurement for your TA.

$I_{in} =$ _____   TA Verification: _____

Post-lab questions: Answer the questions listed below on a separate piece of paper. Make sure that your handwriting is legible! When you are finished, staple everything together and turn in this completed lab packet to your TA.

1. Did your measured I_{in} (in-lab questions 7 and 8) fall within the range you calculated for I_{in} based on the tolerances of your resistors (in-lab questions 5 and 6)? If not, why not?

2. What are other sources of error (other than resistor tolerance) in your measured values for I_{in}?

LAB   4

Voltage Division

Purpose: In this lab, you will work with a set of design constraints to design, draw, build, and test voltage divider circuits.

Pre-lab: A simple voltage divider is shown in Figure 4.1. The source voltage, V_{in}, is *divided* between the two resistors, R_1 and R_2. In this configuration, V_{out} is always a fraction of V_{in}.

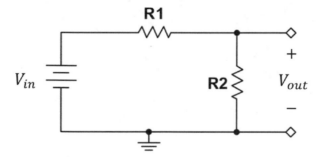

Figure 4.1: Basic voltage divider.

1. For the circuit shown in Figure 4.1, choose R_1 and R_2 to meet the following constraints:

   - The source supplies no more than 0.5 mW when $V_{in} = 10$ V.

   - $V_{out} \approx 0.4 V_{in}$.

   Remember what you learned about design problems in the previous lab: use all of the information you are given to choose R_1 and R_2, and be sure to choose standard resistor values (shown in Appendix D).

   Show your work in the work area, and enter your resistor values here:

   $R_1 = $ _____ $R_2 = $ _____

Work area:

2. Assuming your resistors have a tolerance of ±5%, calculate the possible range for V_{out}. Show your work in the work area, and enter your answer here:

_____ ≤ V_{out} ≤ _____

Work area:

3. Verify that even if both resistors are at their extreme values, you still will not exceed the maximum wattage level. Show your calculations in the work area.

Work area:

In the lab: Design a voltage divider to meet the following constraints:

- The source supplies no more than 800 μW when $V_{in} = 5$ V.
- $V_{out} \approx 0.7V_{in}$.
- You have the following four resistors available: 10 kΩ, 10 kΩ, 10 kΩ, and 20 kΩ (you may have to combine resistors in parallel or series to achieve the resistance that you need).

1. Draw your circuit and show your calculations in the work area.

Work area:

2. Verify that even if all of the resistors are at their extreme values, you will not exceed the maximum wattage level. Show your calculations here.

Work area:

3. Wire your circuit on your breadboard, measure V_{out}, and demonstrate your measurement for your TA.

Measured V_{out} = _____   TA Verification: _____

4. Calculate the expected theoretical V_{out} assuming all of your components are ideal. Show your calculations in the work area.

Calculated V_{out} = _____

Work area:

5. What is the tolerance range for each of your resistors? Using these tolerances, calculate the predicted range for your measured V_{out}. Enter your answer here, and show your work in the work area.

_____ $\leq V_{out} \leq$ _____

Work area:

Post-lab questions: Answer the questions listed below on a separate piece of paper. Make sure that your handwriting is legible! When you are finished, turn in this completed lab packet to your TA.

1. Did your measured V_{out} fall within the range you calculated for V_{out} based on the tolerances of your resistors? If not, why not?

2. What are other sources of error (other than resistor tolerance) in your measured value for V_{out}?

3. What could you do to narrow the range of possible values for V_{out}, i.e., to reduce error in this circuit?

LAB 5

Voltage Division Strikes Back

Purpose: In this lab, you will design a voltage divider circuit to provide multiple, adjustable voltage inputs to an external circuit.

Pre-lab: Figure 5.1 shows a parallel voltage divider, which, as its name suggests, is several voltage dividers connected in parallel. This is one way to use a single power supply to create multiple node voltages.

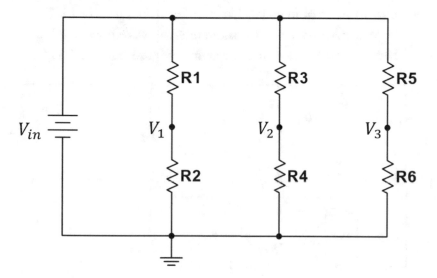

Figure 5.1: Basic parallel voltage divider.

1. Using your knowledge of voltage dividers and the basic circuit template shown in Figure 5.1, design a parallel voltage divider where:

   - $V_{in} = 4$ V,
   - $V_1 \approx 3$ V, $V_2 \approx 2.5$ V, and $V_3 \approx 2$ V, and
   - You have one of each of the following resistors available: 1 kΩ, 2.7 kΩ, 3.3 kΩ, 4.3 kΩ, 7.5 kΩ, and 8.2 kΩ.

   Complete your design, and show your work on a separate piece of paper to submit with this lab. Make sure your calculations are neat and easy to follow. After designing, draw and simulate the circuit in Multisim. Print out the circuit to submit with this lab. The printout must contain indicators (voltage probes, voltmeters, etc.) that show the correct node voltages. Also, the nodes V_1, V_2, and V_3 should be labeled in your drawing.

2. Next, design a parallel voltage divider with adjustable node voltages that will each have a load attached. A template for this circuit design can be seen in Figure 5.2. The design constraints are as follows.

- $V_{in} = 8$ V.
- V_1, V_2, and V_3 should all be able to vary from zero to one volt.
- You have one of each of the following resistors available: 5.6 kΩ, 18 kΩ, and 24 kΩ.
- You have three 10 kΩ potentiometers.
- The load attached to node 1 has an internal resistance of 1 kΩ.
- The load attached to node 2 has an internal resistance of 6.2 kΩ.
- The load attached to node 3 has an internal resistance of 3.3 kΩ.

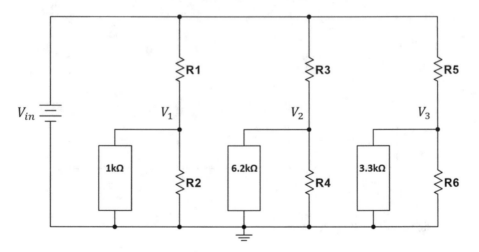

Figure 5.2: Adjustable voltage divider.

As you can see from the design constraints, you will need to use potentiometers in your circuit design. A *potentiometer* is a type of variable resistor that has three terminals and an adjustable dial. (The schematic symbols for a potentiometer and variable resistor are shown in Figure 5.3.) You can use a potentiometer as a simple variable resistor by using only the middle terminal and one of the outside terminals (leaving the other outside terminal unconnected). In this configuration, when you turn the dial, the resistance between the two terminals being used varies between zero and the maximum resistance of the device. It is important to note that potentiometers are most sensitive when operating in their mid-range and not at their extreme values. See Appendix E for more information on potentiometers.

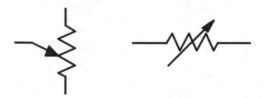

Figure 5.3: Schematic symbol for potentiometer (left) and standard variable resistor (right).

With this in mind, complete your circuit design and show your work on a separate piece of paper to submit with this lab. Make sure your calculations are neat and easy to follow. After designing your circuit, draw and simulate it in Multisim (potentiometers can be found in Multisim under Group: Basic, Family: POTENTIOMETER). Print out the circuit to submit with this lab. The printout must contain indicators (voltage probes, voltmeters, etc.) that display node voltages in the correct range. Also, the nodes V_1, V_2, and V_3 should be labeled in your drawing.

In the lab: Sometimes devices wear out, break, or are used incorrectly, resulting in a short circuit. This can result in very high currents that can be detrimental to the rest of the circuit. One way to protect the circuit from a short is to use a dummy resistor. *Dummy resistors* are low-resistance components usually placed in series with a supply (or other delicate device). One example of this is the 100 Ω resistor in Figure 5.4. If R_1 and R_2 were both shorted, this new dummy resistor would protect the source, V_{in}, from also being shorted and damaged.

Build the second circuit you designed in the pre-lab, with the added protection of a dummy resistor, as shown in Figure 5.4. Measure the voltages at nodes V_1, V_2, and V_3, and verify that you can vary each of their values between zero and one volt.

If your circuit is not working, try using the tips in Appendix C to troubleshoot your circuit. When each of the node voltages is showing the correct reading, show your working circuit to your TA.

TA Verification: _____

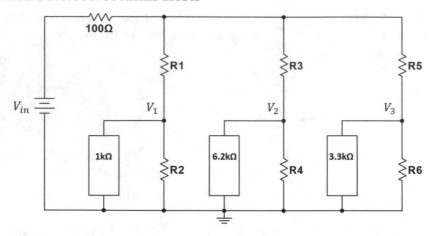

Figure 5.4: Protected parallel, adjustable voltage divider.

Post-lab questions: Answer the questions listed below on a separate piece of paper. Make sure that your handwriting is legible! When you are finished, staple everything together, and turn in this completed lab packet to your TA.

1. By how much did the node voltages you measured in the lab differ from the expected values you saw in Multisim? What do you think accounts for this difference?

2. What would happen if the 100 Ω resistor was 1 kΩ?

3. Why are the potentiometers where they are in the circuit? For example, why can't R_1 be a potentiometer?

4. Would it be possible to just use the potentiometers as voltage dividers, i.e., to remove R_1, R_3, and R_5 from the design? What would this circuit look like?

L A B   6

Temperature Indicator–Part 1: Wheatstone Bridges

Purpose In this lab, you will use a thermistor in a Wheatstone bridge to design the front-end of a temperature-change indicator circuit.

Pre-lab Ultimately, in this two-part lab, we want to design a circuit that will alert us when the ambient temperature has increased by more than 15°F. Part 1 of this lab will concentrate on one way to measure a change in temperature using a thermistor in a configuration called a Wheatstone bridge.

A *negative temperature coefficient* (NTC) *thermistor* is a sensor whose resistance decreases as its temperature increases. Its name comes from combining the words *thermal* and *resistor*. Thermistor values are given in terms of a reference temperature, which is usually 25°C. So, a 10 kΩ thermistor has a resistance of 10 kΩ at 25°C. A thermistor's resistance vs. temperature curve is nonlinear, and you need to consult the device's individual data sheet for exact specifications. The schematic symbol for a thermistor is shown in Figure 6.1, and the data sheet for a thermistor is provided in Appendix F.[1]

Figure 6.1: Thermistor schematic symbol.

A *Wheatstone bridge* is a combination of two voltage dividers. Bridge circuits are traditionally drawn in a diamond shape, and a standard Wheatstone bridge configuration, shown in Figure 6.2, uses four resistive elements in a parallel voltage divider. Of the four resistors in the bridge, two (R_1 and R_3) are known and fixed, one (R_4) is a variable resistor (like a potentiometer), and one (R_2) is an unknown resistance. This unknown resistance is frequently a sensor whose resistance varies with some outside factor, such as temperature, strain, or force. The idea is to connect a voltmeter between V_1 and V_2 and adjust the vari-

---

[1] Appendix F includes temperature vs. resistance data for a Dale/Vishay 01C1002KP Thermistor. If your lab uses a different thermistor, consult your TA for the correct data.

able resistor until the meter shows zero volts between those two nodes; at this point the bridge is *balanced*.

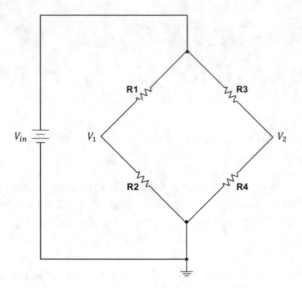

Figure 6.2: Wheatstone bridge.

1. For the circuit in Figure 6.2, express the voltages at V_1 and V_2 in terms of R_1, R_2, R_3, R_4, and V_{in}. Show your work in the work area.

   Work area:

2. What conditions would need to hold for $V_2 - V_1 = 0$? (Hint: solve for R_4.)

Work area:

3. Assume that the starting temperature is 25°C, and, according to the thermistor's data sheet, its resistance is 10 kΩ ± 10%. Assume that $R_1 = R_3 = 1$ kΩ ± 5%, R_4 is a 10 kΩ potentiometer, and $V_{in} = 10$ V. In order for $V_2 - V_1 = 0$, what is the possible range of values for R_4? In other words, what range of values is needed for R_4 to be able to balance the bridge?

Work area:

There is clearly a problem! If the resistance of the thermistor is above 10 kΩ and it is placed in a Wheatstone bridge with a 10 kΩ potentiometer (as configured in Figure 6.2 and the above calculations), the potentiometer may not be able to balance the bridge. One solution to this is to place an additional resistor, say 5.1 kΩ, in series with the potentiometer. By doing this, we ensure that the potentiometer is at approximately half of its maximum value when balancing the bridge.

4. Now assume that, after balancing the bridge in part 3 above, the ambient temperature decreases to 15°C. Assume ideal components (i.e., that $R_1 = R_3$ equal exactly 1 kΩ and R_2 equaled 10 kΩ when you balanced the bridge). Calculate the voltage that will now appear on a meter placed between V_1 and V_2. (*Hint:* Only R_2 varies with temperature; the other three resistances have not changed!)

Work area:

In the lab Using $V_{in} = 10$ V, a 10 kΩ thermistor, a 10 kΩ potentiometer, and any resistors in the lab, design a Wheatstone bridge like the one shown in Figure 6.2. Write the values you've chosen in your design in the spaces below.

$R_1 = $_____ $R_3 = $_____

$R_2 = $_____ $R_4 = $_____

1. Build and test the circuit you designed. Connect the multimeter to measure the voltage difference $V_2 - V_1$, and then vary R_4 (twist the potentiometer's dial) to balance the bridge. Once you think your circuit is working, balance the bridge and call your TA over to help you heat up the thermistor.

TA Verification: _____

2. What is the ambient temperature in the lab, and what is the thermistor's resistance at this starting temperature?

   (a) Ambient temperature=_____

   (b) R_2=_____

3. Once your bridge was balanced and you heated up the thermistor with the lamp, what was the maximum value you saw for $V_2 - V_1$?

   (a) Lamp maximum $V_2 - V_1$=_____

   (b) What temperature difference does this correspond to?

   Work area:

4. Once your bridge was balanced and you heated up the thermistor with the heat gun, what was the maximum value you saw for $V_2 - V_1$?

   (a) Heat gun maximum $V_2 - V_1 =$ _____

   (b) What temperature difference does this correspond to?

      Work area:

Post-lab questions Answer the questions listed below on a separate piece of paper. Make sure that your handwriting is legible! When you are finished, staple everything together and turn in this completed lab packet to your TA.

   1. If an indicator light required at least 2.2 V to turn on, and we attached the light between V_2 and V_1, how large a temperature increase would be required for the light to turn on? Assume the bridge was balanced and the ambient temperature was 25°C to start.

2. In addition to ensuring that we would be able to balance our bridge, why was it a good idea for the potentiometer's resistance to be approximately 5 kΩ when balancing the bridge?

LAB 7

Operational Amplifiers and Thevenin Equivalent Circuits

Purpose: In this lab, you will learn the basics of operational amplifiers, Thevenin equivalent circuits, and designing a circuit for maximum power transfer.

Pre-lab Part 1: Op amps: Contrary to what their schematic symbol suggests (see Figure 7.1), operational amplifiers (op amps) are not actually shaped like triangles! Op amps are integrated circuits (IC), which are devices made up of many individual electronic components that are "printed" on a semiconducting wafer (or "chip"). This single chip can contain only a few transistors, resistors, and capacitors or, in the case of microprocessors, can contain millions of components. The op amps we're using in our lab have input and output pins that serve as connections for mounting the IC on a breadboard.

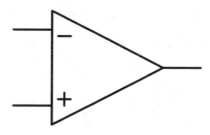

Figure 7.1: Schematic symbol for an op amp.

1. Find the specification sheet (also known as a data sheet or spec sheet) for the LM741 op amp, which is the IC we will be using in this lab. This op amp has eight pins. Identify which pins represent the inverting input, the non-inverting input, and the output for the op amp.

   - inverting input = pin number _____
   - non-inverting input = pin number _____
   - output = pin number _____

These three pins should be familiar to you, but what is the purpose of the other five of the eight pins? One pin is empty, and two pins are labeled as "offset" and "offset null." The function of the offset pins is beyond the scope of this lab.

The remaining two pins are labeled as V_{cc+} and V_{cc-}. They power the op amp and define the linear range of the output voltage. V_{out} cannot be greater than V_{cc+} or less than V_{cc-}, and to approach these limits is to run the risk of your op amp no longer operating in the linear range. If the op amp is operating outside of the linear range, it is *saturated* and will not perform as expected. A safe operating range, then, is shown in Equation 7.1.

$$(V_{cc-} + 2) \leq V_{out} \leq (V_{cc+} - 2). \tag{7.1}$$

These extra two volts give you a cushion so that you don't get too close to saturation. The IC's specification sheet lists the recommended operating conditions as

$$\begin{aligned} 5 \text{ V} \leq V_{cc+} &\leq 15 \text{ V} \\ -15 \text{ V} \leq V_{cc-} &\leq -5 \text{ V}. \end{aligned} \tag{7.2}$$

2. With this information, what is the minimum and maximum output voltage you could reasonably expect this op amp to produce?

$$\underline{\hspace{3cm}} \leq V_{out} \leq \underline{\hspace{3cm}}$$

Which pins represent V_{cc+} and V_{cc-}?

- V_{cc+} = pin number _____
- V_{cc-} = pin number _____

3. Design a unity gain (gain = $\left| \dfrac{V_{out}}{V_{in}} \right|$ = 1) inverting amplifier with the following constraints:

- You have a single voltage source, $V_{in} = 4$ V.
- You have available two 5 kΩ, two 2 kΩ, and one 1 kΩ resistors.
- You have one three-terminal ideal op amp.

(*Note:* Not all components will need to be used.) Show your work in the work area. Then draw and simulate your circuit in Multisim. (Op amps can be found under Group: Analog, Family: ANALOG_VIRTUAL. Choose OPAMP_3T_VIRTUAL.) Include a voltage probe or voltmeter that shows that $V_{out} = -V_{in}$. Print out the drawing to submit with this lab.

Work area:

4. We want to add a load to the circuit shown in Figure 7.2, and we need the voltage across the load (V_{load}) to be equal to V_N. First, calculate the voltage V_N before the load is attached.

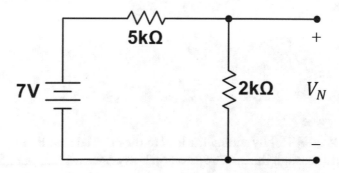

Figure 7.2: Ye olde voltage divider.

Enter your answer here, and show your work in the work area.

$V_N =$ _____

Work area:

5. Assume that the load can be modeled as a 10 kΩ resistor. Calculate what the voltage across the load (V_{load}) would be if it were simply attached in parallel to the 2 kΩ resistor. Enter your answer here, and show your work in the work area.

$V_{load} =$ _____

Work area:

Clearly, $V_{load} \neq V_N$! By adding the load to the original circuit in this way, we changed the current through the 2 kΩ resistor. So how do we achieve our goal of $V_{load} = V_N$? A unity gain, non-inverting amplifier can act as a buffer, ensuring that none of the current from the original circuit will flow to the load. This is called a *buffer amplifier* and will allow you to add the 10 kΩ load to the original circuit and achieve your goal of $V_{load} = V_N$.

6. Add a buffer amplifier to your circuit, and draw your design in the work area. Show calculations to verify that $V_{load} = V_N$.

Work area:

Pre-lab Part 2: Thevenin equivalent circuits: *Thevenin's theorem* states that any two-terminal, linear circuit can be replaced by an equivalent circuit consisting of a single voltage source in series with a single resistor. These circuits are referred to as Thevenin equivalent circuits. The *maximum power transfer theorem* states that the maximum amount of power is transferred to the load when the load resistance is equal to the Thevenin resistance. See Appendix G for information on how to find a Thevenin equivalent circuit.

1. Thevenin equivalents can be useful when designing a circuit with a pre-built front end or using a circuit that you cannot change, such as one soldered to a printed circuit board (PCB). One such front end is provided in Figure 7.3, and, as you probably have

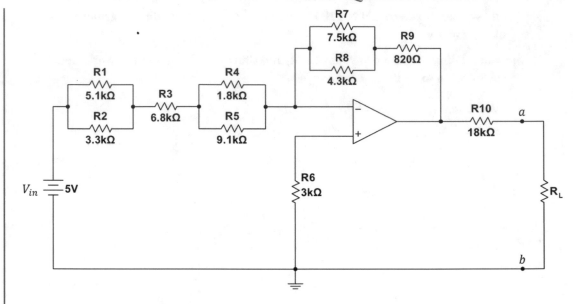

Figure 7.3: Complex circuit you don't want to solve more than once.

guessed, your next task is to find a Thevenin equivalent for this circuit. Show your work on a separate piece of paper to include with this lab.

2. Draw your Thevenin equivalent circuit here:

3. Use your Thevenin equivalent circuit to complete the following calculations:

(a) What should the load resistance be in order to receive maximum power from this circuit?

For max power, $R_L =$ _____

(b) What is the power given to the load when power is maximized?

Max power to load, $P_{max} =$ _____

(c) Now, for each load resistance shown in Table 7.1, calculate how much power each would receive. Also calculate what the voltage across and the current through each load would be, and fill in Table 7.1. Show your calculations in the work area.

Table 7.1: Load resistance values for Thevenin equivalent circuit of Figure 7.3

Load Resistance	Load Voltage (V) (Calculated)	Load Current (μA) (Calculated)	Load Power (μW) (Calculated)
$R_L = 18$ kΩ			
$R_L = 1$ kΩ			
$R_L = 9.1$ kΩ			
$R_L = 43$ kΩ			
$R_L = 68$ kΩ			

Work area:

In the lab: Your TA will explain how to breadboard the op amp and how to choose appropriate values for V_{cc+} and V_{cc-}. Build the circuit in Figure 7.3, with the load resistor, R_L, removed. Once built measure the voltage across the open.

$$V_{oc} = V_{ab} = \underline{\hspace{3cm}}$$

One at a time, attach each load resistance given in Table 7.2 between a and b in your circuit, and measure the voltage across and the current through each one. Fill in Table 7.2 with your measured values, and calculate the power being delivered to each load resistor. While doing this, demonstrate to your TA that your circuit is working correctly.

TA Verification: \underline{\hspace{4cm}}

Table 7.2: Measured load resistance values for circuit in Figure 7.3

Load Resistance	Load Voltage (V) (Measured)	Load Current (μA) (Measured)	Load Power (μW) (Calculated)
R_L = 18 kΩ			
R_L = 1 kΩ			
R_L = 9.1 kΩ			
R_L = 43 kΩ			
R_L = 68 kΩ			

Post-lab questions: Answer the questions listed below on a separate piece of paper. Make sure that your handwriting is legible! When you are finished, turn in this completed lab packet to your TA.

1. How closely did your measured V_{oc} match your calculated V_{th}?

2. Did your measured load voltages match the values you calculated in the pre-lab? How close were they?

3. Was using the Thevenin equivalent circuit worth it?

LAB 8

Temperature Indicator–Part 2: Difference Amplifiers

Purpose: In this lab, you will use a thermistor in a Wheatstone bridge, an LED, and a difference amplifier to design a temperature-change indicator circuit.

Pre-lab: We want to design a circuit that will alert us when the ambient temperature has increased by more than 15°F. Design a circuit that meets the following criteria.

- Assume that the starting ambient temperature in the room is 22.2°C.
- You have a single voltage source, $V_{in} = 10$ V.
- When the ambient temperature exceeds 30°C, an indicator light will turn on.

You have the following components available:

- a single red LED, whose turn-on voltage is approximately 2.2 V,
- any standard resistors provided in lab,
- up to three LM741 op amps,
- one 10 kΩ potentiometer, and
- one thermistor, rated 10 kΩ ± 10% at 25°C (see data sheet in Appendix F).

An *LED* (light-emitting diode) is an electronic component that allows current to flow in only one direction, and, when it receives more than a certain minimum voltage between its two terminals, it will turn on. LEDs come in different colors, which have different turn-on voltages. The schematic symbol and polarity for an LED are shown in Figure 8.1, and Figure 8.2 shows a photo of an LED. In this lab, we are treating the LED as a load that requires 2.2 V to operate.

Figure 8.1: Schematic symbol for an LED.

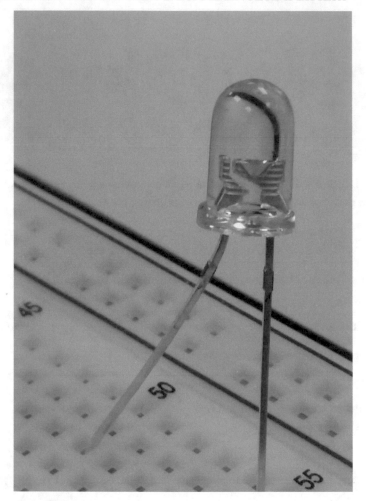

Figure 8.2: An LED striking a pose [3].

In Lab 6 (Part 1 of this lab), we saw that by attaching a thermistor to a Wheatstone bridge (see Figure 8.3), we could measure a difference in temperature as a difference in voltage between nodes V_1 and V_2.

We also saw in the Lab 6 post-lab questions that in order for $V_2 - V_1 = 2.2\,\text{V}$, the ambient temperature would have to increase by more than $40°\text{F}$! So, if the LED requires at least $2.2\,\text{V}$ to turn on and we want the LED to light when the temperature changes by $\sim 15°\text{F}$, we cannot merely connect the LED between V_1 and V_2. We need to amplify the voltage difference between V_1 and V_2.

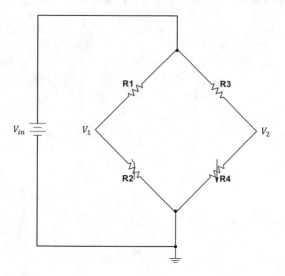

Figure 8.3: Wheatstone bridge.

1. Why not use an op amp configured as a difference amplifier? The difference amplifier shown in Figure 8.4 has an output that is proportional to the voltage difference between the two inputs. Calculate V_{out} in terms of V_1, V_2, R_5, and R_6, and show your calculations in the work area.

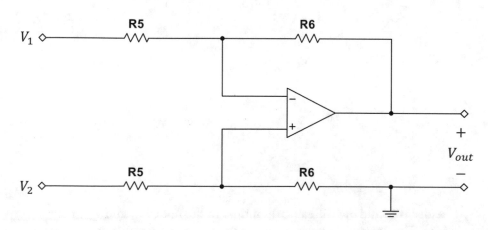

Figure 8.4: Difference amplifier.

Work area:

The idea is to add our difference amplifier to the nodes V_1 and V_2 in our Wheatstone bridge. Then, we can balance the bridge at our starting ambient temperature, and any changes in temperature will result in a non-zero voltage difference between V_1 and

V_2, which will be amplified by the difference amplifier. If the amplified difference is large enough, the LED will light.

2. Using the values for R_1 through R_4 from your Wheatstone bridge design in Lab 6, calculate what the voltage $V_2 - V_1$ will be when the ambient temperature reaches 30°C (recall the starting ambient temperature is 22.2°F).

Work area:

3. Use this information to choose values for R_5 and R_6, and write your chosen resistor values in the spaces provided. In other words, choose R_5 and R_6 so that the $V_2 - V_1$ you just calculated will result in a $V_{out} = 2.2$ V.

$R_1 =$_____ $R_4 =$_____

$R_2 =$_____   _____ $R_5 =$_____

$R_3 =$_____ $R_6 =$_____

Work area:

4. If we add the difference amplifier directly to V_1 and V_2, we end up with the circuit in Figure 8.5. It seems like this circuit should work, but it won't! Recall from Lab 7 what happens when we add a load directly to a voltage divider. In Lab 7, our load was just an additional resistor, but in Figure 8.5, we are adding a load in the form of a difference amplifier to nodes V_1 and V_2. Before, when the Wheatstone bridge was by itself, R_3 and R_4 had the same current through them. Now, some of that current is diverted to R_5 and R_6. In fact, R_4 is now effectively in parallel with the R_5 - R_6 combination!

How can we fix this? We could re-do our entire analysis of the circuit shown in Figure 8.5, choosing new values for all the resistors that would give us the output we want and satisfy all of our design criteria. Or we could keep our original resistor values and figure out a way to isolate the Wheatstone bridge half of the circuit from the difference amplifier so that no current is shared between them. Use the work space

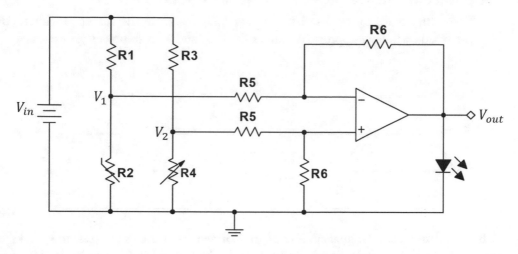

Figure 8.5: Temperature-change indicator circuit using a Wheatstone bridge with a difference amplifier.

to draw a new circuit, either with all new resistor values or with the two parts of the circuit isolated from each other.

Work area:

Write the values for each resistor in your final design in the spaces provided. Also, verify that you have appropriate values for V_{cc+} and V_{cc-} to power the op amp.

$V_{cc+} =$ _____ $V_{cc-} =$ _____

$R_1 =$ _____ $R_4 =$ _____

$R_2 =$ _____ $R_5 =$ _____

$R_3 =$ _____ $R_6 =$ _____

In the lab: Build and test your final circuit design. Connect a voltmeter to measure $V_2 - V_1$ and vary R_4 (twist the potentiqmeter's dial) to balance the bridge. Then, vary R_4 until the LED turns on. Once you think your circuit is working, balance the bridge and call your TA over to help you heat up the thermistor.

TA Verification: _____

Post-lab questions: Answer the questions listed below on a separate piece of paper. Make sure that your handwriting is legible! When you are finished, staple everything together and turn in this completed lab packet to your TA.

1. If we wanted the LED to turn on when the temperature *decreased* by a certain amount, what could we do?

2. What are some of the drawbacks of this circuit design? Was your circuit sufficiently sensitive and accurate? What could you do to improve the accuracy?

LAB 9

What's my Grade? (Part 1)

Purpose: In this lab, you will build an analog computer using a parallel voltage divider and a summing amplifier to calculate a student's final grade in a class.

Pre-lab: As the end of the semester nears, students begin worrying more and more about their final grades. One student in particular is concerned about her circuits class: she wants to know what grade she needs on her final exam to get an "A" in the class. She has decided to demonstrate her EE prowess by building a circuit that will take her average scores on various assignments as inputs and will output her final percentage in the class. The first part of this lab will focus on calculating her final score; the second part of this lab will focus on designing a way to allow multiple, adjustable inputs.

1. The first thing our student needs to do is understand how to compute her final grade in her circuits class. According to the course syllabus, her grade will be a combination of homework assignments, quizzes, two exams, and a final exam, weighted as follows:

Homework:	15%
Quizzes:	15%
Exam I:	20%
Exam 2:	20%
Final Exam:	30%

   Her final grade will be a weighted sum of her average homework score, her average quiz score, and her scores on each exam. Assume that, to date, her homework average is 82%, her quiz average is 78%, she received a 76% on the first exam, and an 82% on the second exam. If she receives an 80% on her final exam, what will her final score be in the class? Show your work in the work area.

Work area:

2. How can she use common circuit components to do this calculation for her? Consider the three-input summing amplifier shown in Figure 9.1. This summing amplifier's output voltage is a weighted sum of the input voltages, and the weighting depends on the ratio of the feedback resistor (R_f) to the resistance at each input (R_1, R_2, and R_3). Calculate V_{out} in terms of V_1, V_2, V_3, R_1, R_2, R_3, and R_f, and show your work in the work area.

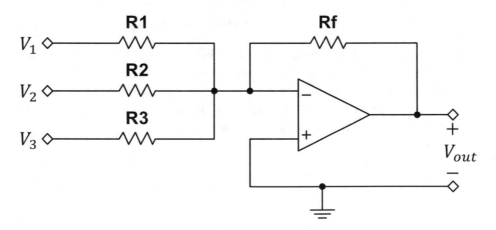

Figure 9.1: Three-input summing amplifier.

Work Area:

3. With all of this in mind, design a summing amplifier that will calculate our student's final grade. Make sure the circuit meets the following criteria:

- Assume the following grade weighting for her circuits class:
  - Homework: 15%
  - Quizzes: 15%
  - Exam I: 20%
  - Exam 2: 20%
  - Final Exam: 30%
- Assume that her professor is using a standard grading scale for the class:
  - 90–100% = A
  - 80–89% = B
  - 70–79% = C
  - 60–69% = D
  - <60% = F
- Assume each input voltage (V_1 through V_5) is between zero and one and represents her average grades as follows:
  - V_1 = Homework average
  - V_2 = Quiz average

- V_3 = Exam 1 score
- V_4 = Exam 2 score
- V_5 = Final Exam score

- The output, V_{out}, is a positive voltage between zero and one and represents her final score in the class.

You have the following available components:

- any standard resistors provided in the lab,
- up to two LM741 op amps, and
- five voltage supplies.

Show your work in the work area. Once your design is complete, draw your circuit in Multisim. Choose any values you like for V_1 through V_5, and simulate your circuit. Print out your drawing to include with this lab, and make sure that the input and output voltage values are visible in your drawing.

Work area:

In the lab: Build the circuit you designed in the pre-lab, leaving the input voltages off (disconnected). Enter the values you've chosen for each of the resistors in your design, and choose appropriate values for V_{cc+} and V_{cc-}.

$V_{cc+} = $ _____ $V_{cc-} = $ _____

$R_1 = $ _____ $R_5 = $ _____

$R_2 = $ _____ $R_f = $ _____

$R_3 = $ _____   Other $R = $ _____

$R_4 = $ _____   Other $R = $ _____

1. Since each lab station does not have five voltage supplies, we will be using a single input voltage supply and the theory of superposition to compute our total output voltage. Using the values in Table 9.1 as your inputs, connect the voltage source to the first node (V_1) and measure the output voltage. Repeat for each input, and write the values you measured for V_{out} in Table 9.1. When finished, compute the total V_{out} for each column.

Table 9.1: Measured V_{out} for each input

Input (V)	V_{out} (V)	Input (V)	V_{out} (V)
$V_1 = 0.60$		$V_1 = 0.72$	
$V_2 = 0.65$		$V_2 = 0.63$	
$V_3 = 0.50$		$V_3 = 0.68$	
$V_4 = 0.65$		$V_4 = 0.71$	
$V_5 = 0.75$		$V_5 = 0.84$	
Total:		Total:	

Show your working circuit to your TA.   TA Verification: _____

2. Since our student wants to figure out how her final grade will affect her score, she has filled in Table 9.2 with her current homework and quiz averages, along with her scores on the first two exams. Measure the output voltage for each input, and, when finished, compute the total V_{out} for inputs V_1 through V_4.

Table 9.2: Our student's current scores

Input (V)	V_{out} (V)
$V_1 = 0.92$	
$V_2 = 0.90$	
$V_3 = 0.88$	
$V_4 = 0.82$	
Total:	

3. Now measure V_{out} for the range of potential final exam scores shown in Table 9.3. Add each measured V_{out} due to V_5 alone to the total from Table 9.2 to calculate the final grade she would receive for each final exam score.

4. Since you will be completing the second half of this lab next week, use the rest of your lab time to start working on next week's pre-lab. It would be most efficient if you completed the full design with your lab partner(s) so that you and your partner(s) have all of the same resistor values coming into next week's lab period. You will need the values from this week's design (R_1 through R_5 and R_f) for next week, so be sure to copy them down.

Table 9.3: Potential range of final exam scores and their corresponding effect on final grade

Final Exam Score (%/100)	V_{out} (V)	Table 9.2 Total + V_{out} = Final Grade
$V_5 = 1.0$		
$V_5 = 0.90$		
$V_5 = 0.80$		
$V_5 = 0.70$		
$V_5 = 0.60$		

Post-lab questions: Answer the questions listed below on a separate piece of paper. Make sure that your handwriting is legible! When you are finished, staple everything together and turn in this completed lab packet to your TA.

1. Based on the numbers in Table 9.3, what is the minimum grade our student needs on her final exam to receive an "A" in her circuits class?

2. Using the values given in Table 9.2, calculate, theoretically, what minimum grade our student needs on her final exam to receive an "A" in her circuits class.

3. How could you make your circuit more accurate?

LAB 10

What's my Grade? (Part 2)

Purpose: In this lab, you will build an analog computer with adjustable inputs to calculate a student's final grade in a class, using a parallel voltage divider and a summing amplifier.

Pre-lab: A student wants to build an analog computer to calculate her grade in her circuits class. Ideally, she wants be able to calculate the final grade of *any* student in the class with this circuit. Unlike the design in last week's lab, we need a circuit with adjustable inputs and a single voltage source. This way, we could change any of the input grades (i.e., V_1 through V_5) and watch the final grade (i.e., V_{out}) change in real-time. With this final goal in mind, design a circuit that meets the following criteria:

- Assume the following grade weighting for our student's circuits class:
  - Homework: 15%
  - Quizzes: 15%
  - Exam I: 20%
  - Exam 2: 20%
  - Final Exam: 30%
- Assume each input voltage (V_1 through V_5) can vary between zero and one and represents her average grades as follows:
  - $V_1 = $ Homework average
  - $V_2 = $ Quiz average
  - $V_3 = $ Exam I score
  - $V_4 = $ Exam 2 score
  - $V_5 = $ Final Exam score
- The output, V_{out}, is a positive voltage between zero and one and represents the final score in the class.
- You have one available voltage source, $V_{in} = 2$ V.

You have the following available components:

- any standard resistors provided in the lab,

- up to two LM741 op amps, and

- up to five 10 kΩ potentiometers.

You can use your summing amplifier design from last week's lab; however, you need to design a front-end that will give you multiple, adjustable node voltages. (*Hint:* review Lab 5!) These adjustable node voltages should vary between zero and one volt and will be connected directly to your summing amplifier inputs V_1 through V_5. This will allow you to input any student's scores and see a final grade at the output.

Show your design work here and in the work area. Once your design is complete, draw your circuit in Multisim. Choose any values you like for V_1 through V_5 and simulate your circuit. Print out your drawing to include with this lab, and make sure that the input and output voltages are visible in your drawing.

Work area:

In the lab: Build the circuit you designed in the pre-lab. You should be able to see a change in V_{out} when you vary each input potentiometer. When you think your circuit is working, put in some test cases for V_1 through V_5 (e.g., the values you chose when you simulated your Multisim drawing) and verify that you are getting the correct V_{out}.

1. Since last week, our student turned in two homework assignments and took a quiz. Her new averages are shown in Table 10.1. Calculate, theoretically, what grade she needs on her final exam to receive an "A" in the class. Show your work in the work area and enter your calculated answer here.

   Calculated final exam score needed for an "A" in the class: _____

Table 10.1: Our student's current scores

Homework = 0.92
Quizzes = 0.91
Exam I = 0.88
Exam II = 0.82

Work area:

2. Now use your circuit to determine what grade she needs on her final exam to receive an "A" in the class. Write your measured value here: _____

3. Demonstrate your working circuit to your TA. Your TA will give you a final exam grade to input into your circuit. Write that grade in the space below, and also write your circuit's final grade answer below.

- Final exam score from TA: _____
- Final grade in class (V_{out} measured from circuit): _____

                              TA Verification: _____

Post-lab questions: Answer the questions listed below on a separate piece of paper. Make sure that your handwriting is legible! When you are finished, staple everything together and turn in this completed lab packet to your TA.

1. What if you don't want to measure the output voltage each time you use the circuit? Instead you want an indicator light to show whether or not your input grades result in the desired final grade. Sketch a circuit that you could attach to the output of your summing amplifier that would turn on an LED when the final grade is above a 70%. (Recall that a red LED has a turn-on voltage of 2.2 V.)

LAB   11

Fun with LEDs

Purpose: In this lab, you will design a phase-shifting circuit to make four LEDs blink in a repeating sequence.

Pre-lab: In this lab we will be working with capacitors and an AC input for the first time. AC stands for *alternating current* and refers to an input current or voltage that varies between positive and negative values. Examples of AC waveforms include sinusoids and square waves. See Figures 11.1 and 11.2 for examples. In the lab, in order to supply a circuit with an AC signal, we use a function generator, which can produce a variety of waveforms,

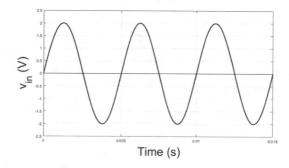

Figure 11.1: AC sinusoidal signal.

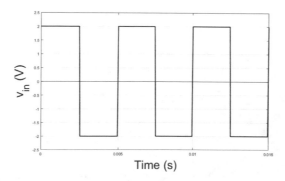

Figure 11.2: AC square signal.

including sinusoids, square waves, and triangle waves. In order to view all of the varying voltages and currents in the circuit, we will use an oscilloscope.

1. One interesting AC circuit is the differentiator shown in Figure 11.3. For this circuit, derive the equation for the output, $v_{out}(t)$, in terms of $v_{in}(t)$, C, and R. Do this derivation in the time-domain, and show your calculations in the work area.

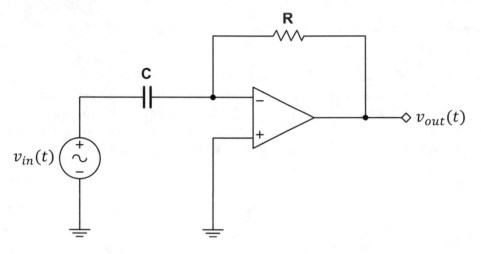

Figure 11.3: Basic differentiator circuit.

Work area:

2. Assume the circuit shown in Figure 11.3 has a $v_{in}(t) = 5\cos(1000 * 2\pi t)$ mV, $C = 10\ \mu F$, and $R = 15\ k\Omega$. Use a frequency-domain technique (such as phasors) to find $V_{out}(j\omega)$ and $v_{out}(t)$, and show your work in the work area.

Work area:

Note that the output of this circuit is the input scaled and shifted by 90°. We can control the amplitude and the phase of the output with this circuit, but we cannot change the frequency. The frequency of the output will be the same as the frequency of the input.

Now draw and simulate this circuit in Multisim. Use a function generator for $v_{in}(t)$, and add an oscilloscope to your drawing. (See Appendix B for details on finding these new components.) Connect both your input and output to the oscilloscope to display both signals together. Print out your drawing to include with this lab, and make sure that the input and output waveforms are displayed in the oscilloscope window.

3. Keeping your differentiator in mind, design a circuit that meets the following criteria.

   • Your circuit has four outputs that are phase-shifted 90° from each other.
   • One LED is attached to each output so that the four LEDs are placed in a row. These LEDs should blink in a repeating sequence, with only one LED lit at a time. This idea is shown graphically in Figure 11.4.

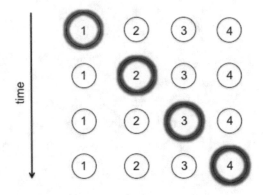

Figure 11.4: Graphic depiction of four sequentially blinking LEDs.

You have the following supplies and components available:

   • four red or green LEDs with turn-on voltages of ≈ 2.2 V,
   • a single sinusoidal voltage source, $v_{in}(t) = 5\cos(2\pi f t)$ V, where $f = 0.2$ Hz,
   • four 10 kΩ potentiometers,
   • any standard resistors that are greater than 100 Ω and are provided in the lab,
   • three LM741 op amps, and
   • any of the following capacitors: 10μF, 0.33μF, 0.03μF, 0.02μF, 0.0075μF, and 0.002μF.

Note that you are not required to use all of the available components and should try to design a working circuit that uses the fewest components possible. Show your work in the work area. Once your design is complete, draw and simulate your circuit in Multisim. (LEDs can be found under Group: Diodes, Family: LED. LED_green and/or LED_red would be the most appropriate to use.) Print out your drawing to include with this lab.

Work area:

In the lab: You will be building the two circuits you analyzed and designed in the pre-lab. Your TA will demonstrate the basics of using a function generator and oscilloscope.

1. First build the differentiator shown in Figure 11.3. Use a triangle wave with an amplitude of 2 V and frequency of 40 Hz as your $v_{in}(t)$ (see Figure 11.5), and use $C = 10\ \mu F$ and $R = 1\ k\Omega$. Calculate the range of output voltages you expect to see, and choose appropriate $V_{cc\pm}$ values. Use the oscilloscope to display both the input and the output.

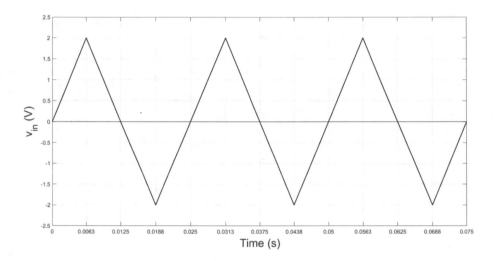

Figure 11.5: Triangle wave to be used as $v_{in}(t)$ for the differentiator.

Demonstrate your working circuit to your TA, and use the template shown in Figure 11.6 to sketch what the output voltage looks like.

TA Verification: _____

2. Add an LED across your output, and replace the resistor in your circuit with a 10 kΩ potentiometer. What happens when you vary the resistance? (Write your answer in the work area.) Show your working circuit to your TA.

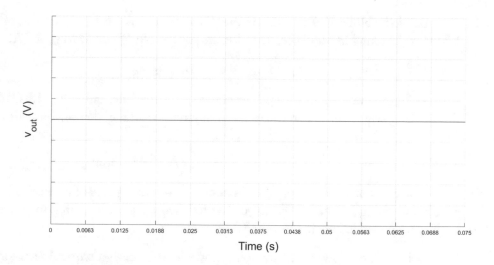

Figure 11.6: Graph template for sketching the differentiator $v_{out}(t)$.

Work area:

TA Verification: _____

3. Now build the blinking-LED circuit you designed in the pre-lab. When all of your LEDs are blinking in sequence, demonstrate your working circuit to your TA.

TA Verification: _____

4. **Bonus:** Figure out how to turn off, or remove, in real-time, individual LEDs from the blinking sequence. When you've got your circuit working, demonstrate it to your TA.

TA Verification: _____

Post-lab questions: Answer the questions listed below on a separate piece of paper. Make sure that your handwriting is legible! When you are finished, staple everything together and turn in this completed lab packet to your TA.

1. If the LEDs had a lower turn-on voltage, would your design work? What kind of adjustments would you make?

2. If you use a square wave as the input to a differentiator, what is the output?

LAB 12

Circuits Dance Party

Purpose: In this lab, you will design a circuit to filter out low-frequency signals and a circuit to filter out high-frequency signals.

Pre-lab: One of the most basic and widely used AC circuits is called a filter. A *filter* is a circuit that either allows or blocks a particular range of frequencies. A *low-pass filter* is a circuit that allows only low frequencies to pass through. Higher-frequency inputs will be attenuated (amplitude is reduced or flattened) or even completely blocked. Low-pass filters are used, for example, as inputs to subwoofers or bass speakers because such speakers cannot effectively reproduce high-frequency sounds. In the same vein, *high-pass filters* are circuits that allow only high frequencies to pass through; lower-frequency inputs will be attenuated or blocked. High-pass filters are used, for example, as inputs to tweeters, to remove low-frequency, bass sounds that may damage these speakers. Both high-pass and low-pass filters are widely used in all areas of audio, image, and other types of signal processing.

1. At their most basic, low-pass and high-pass filters are first-order (RC or RL), AC circuits. (See Figure 12.1.)

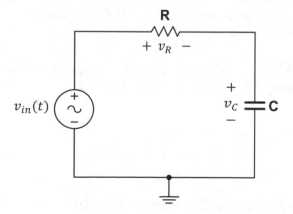

Figure 12.1: A simple RC circuit.

(a) For the circuit in Figure 12.1, use a frequency-domain technique (such as phasors) to derive the equation for the voltage across the capacitor, shown in Equation 12.1. Assume that $v_{in}(t) = A\cos(\omega t)$ V. Show all of your work on a separate piece of paper.

$$v_c(t) \quad = \quad \frac{A}{\sqrt{1 + (\omega RC)^2}}\cos(\omega t + \Phi_1). \qquad (12.1)$$

(b) Now derive the equation for the voltage across the resistor in Figure 12.1, shown in Equation 12.2. Show all of your work on a separate piece of paper.

$$v_R(t) \quad = \quad \frac{\omega RCA}{\sqrt{1 + (\omega RC)^2}}\cos(\omega t + \Phi_2). \qquad (12.2)$$

It is clear from Equations 12.1 and 12.2 that the amplitudes of the voltages across both the capacitor and the resistor depend on the frequency of the input signal.

2. How do you know if a circuit is a low-pass or high-pass filter, and how do you know where the cutoff is? In other words, for a low-pass filter, how do you know which range of low frequencies will pass unchanged?

This information is determined by the *cutoff frequency*, f_c. In a low-pass filter, inputs with frequencies below f_c will be transmitted through the circuit unharmed, and inputs with frequencies above f_c will be attenuated (reduced) or blocked. In a high-pass filter, the opposite is true: inputs with frequencies below the cutoff frequency will be attenuated or blocked, and inputs with frequencies above the cutoff frequency will be allowed to pass.

We can calculate f_c as shown in Equation 12.3, where R is the resistance and C is the capacitance in the RC circuit. Also, recall that $\omega = 2\pi f$ rad/s, so f_c is measured in Hertz (Hz).

$$f_c = \frac{1}{2\pi RC}. \qquad (12.3)$$

3. For the circuit in Figure 12.1, assume that $v_{in}(t) = 10\cos(\omega t)$ V, $R = 100\ \Omega$, and $C = 10\ \mu$F. Use these values to populate Table 12.1, and calculate the cutoff frequency for this circuit.

$f_c = $ _____

Table 12.1: Amplitude values for $v_c(t)$ and $v_R(t)$ at varying input frequencies

f (Hz)	ω (rad/s)	Amplitude $v_c(t)$	Amplitude $v_R(t)$
1			
10			
100			
1000			
10 k			
100 k			
1 M			

You can see that, for very low frequencies, the voltage across the capacitor has nearly the same amplitude as the input. You can also see that capacitor voltage amplitude starts to be attenuated between 10 Hz and 100 Hz. For an input frequency of 10 kHz, the voltage across the capacitor is very small, so high frequencies are being blocked. In the same way, for very high frequencies, the voltage across the resistor has nearly the same amplitude as the input; however, the resistor voltage amplitude starts to be attenuated between 1 kHz and 100 Hz. For an input frequency of 1 Hz, the voltage across the resistor is very small, so low frequencies are being blocked.

This makes sense if you consider that the circuit in Figure 12.1 is an AC voltage divider, and $v_c + v_R = v_{in}$. Think about what is happening in this circuit. If we have a very low frequency input, say a DC input (whose frequency is zero!), the capacitor has time to charge fully, and in a very short time the circuit is in DC steady-state. With the capacitor acting like an open, $v_c = v_{in}$ and $v_R = 0$. Conversely, if we have an input with a very high frequency, the capacitor does not have time to charge at all; its voltage is therefore zero, and $v_R = v_{in}$.

4. Based on the values in Table 12.1, where would you measure the output for a low-pass filter? Where would you measure the output for a high-pass filter?

   - low-pass filter = _____
   - high-pass filter = _____

5. Plot the values from Table 12.1 on the blank template shown in Figure 12.2. Note that the horizontal axis is shown on a log scale.

6. Using Equation 12.3, design an RC circuit with a cutoff frequency of 300 Hz. Assume that the input is $v_{in}(t) = \cos(\omega t)$ V. Be sure to choose standard resistor and

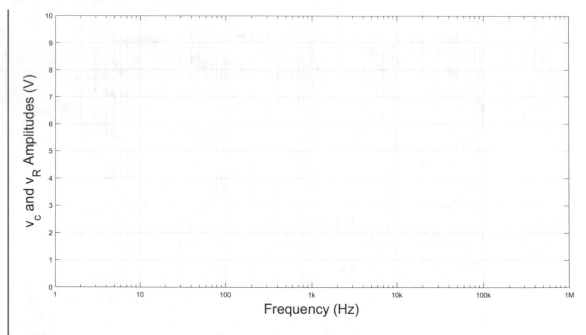

Figure 12.2: Template to graph $v_c(t)$ and $v_R(t)$ amplitudes vs. frequency, f.

capacitor values (standard capacitor values are included in Appendix D). Show your work in the work area and enter your resistor and capacitor values in the spaces provided. (Hint: it is easier to choose a standard capacitor value and then calculate the resistance needed, as you are already familiar with combining resistors to find a needed equivalent resistance.)

For $f_c = 300$ Hz: $R_1 = $ _____ $C_1 = $ _____

Work area:

Use Equations 12.1 and 12.2 to populate Table 12.2. Draw and simulate your circuit in Multisim, using a function generator as the input. Show the waveforms for v_c and v_R using an oscilloscope.

Table 12.2: Amplitude values for $v_c(t)$ and $v_R(t)$ at varying input frequencies ($f_c = 300$ Hz)

Input Frequency f (Hz)	Input Frequency ω (rad/s)	Amplitude $v_c(t)$	Amplitude $v_R(t)$
1			
10			
100			
1000			
10 k			
100 k			
1 M			

7. Using Equation 12.3, design an RC circuit with a cutoff frequency of 3 kHz. Assume that the input is $v_{in}(t) = \cos(\omega t)$ V. Show your work in the work area and enter your resistor and capacitor values in the spaces provided.

For $f_c = 3$ kHz: $R_2 = $ _____ $C_2 = $ _____

Work area:

Use Equations 12.1 and 12.2 to populate Table 12.3. Draw and simulate your circuit in Multisim, using a function generator as the input. Show the waveform for v_c and v_R using an oscilloscope.

Table 12.3: Amplitude values for $v_c(t)$ and $v_R(t)$ at varying input frequencies ($f_c = 3$ kHz)

Input Frequency f (Hz)	Input Frequency ω (rad/s)	Amplitude $v_c(t)$	Amplitude $v_R(t)$
1			
10			
100			
1000			
10 k			
100 k			
1 M			

In the lab: Build both of the circuits you designed in the pre-lab ($f_c = 300$ Hz and $f_c = 3$ kHz) together on the same breadboard, and connect the same input to each. We want both circuits to receive the same input, but we don't want the circuits to be wired together or share any current.

1. Starting with the $f_c = 300$ Hz circuit and using $v_{in}(t) = 2\cos(1000 * 2\pi t)$ V, display both the capacitor voltage and the resistor voltage on an oscilloscope. Vary the frequency of the input and observe what happens to the two output signals. Compare what you are observing with the values you calculated for Table 12.2.

   Repeat this with the $f_c = 3$ kHz circuit and observe the two output signals. Compare what you are observing with the values you calculated for Table 12.3. Demonstrate your working circuit to your TA.

   TA Verification: _____

2. Keeping in mind that both circuits can function as both a high-pass and a low-pass filter, replace your $v_{in}(t)$ with an audio jack connected to an .mp3 player or phone. Where should you place an additional audio jack to best receive a treble-only signal and a bass-only signal? Write your answer in the work area.

Work area:

3. With music as an input and connecting headphones to your output(s), compare the performance of your two circuits. Trying listening to the same song at each of the four outputs.

Post-lab questions: Answer the questions listed below on a separate piece of paper. Make sure that your handwriting is legible! When you are finished, staple everything together and turn in this completed lab packet to your TA.

1. How could you vary the cutoff frequency in real-time? How much could you vary the cutoff frequency (give a range)?

APPENDIX A

Voltage and Current Measurements

Voltage and current are measured differently, and it is important to understand why. Incorrectly measuring these quantities will not only yield confusing readings, it might also result in damage to the multimeter. The following explanation for taking measurements is based on the circuit presented in Figure A.1.

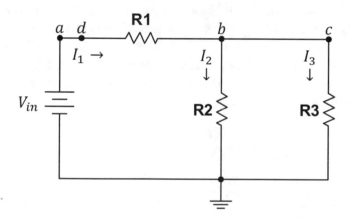

Figure A.1: A simple circuit.

Voltage:

We know that components that are in parallel with each other (e.g., that share the same two nodes) have the same voltage across them. Therefore, when measuring voltage, you are essentially putting the voltmeter in parallel with one or more components.

There are two main types of voltages to measure. One is the voltage across a specific component or source, and measuring it is very straightforward. You only need to put the voltmeter leads on either side of the component. For example, to measure the voltage across R_1, place multimeter's red lead at point a and the black lead at point b. This voltage would be labeled V_{ab} to show which lead was placed at which location in the circuit. If the voltage was labeled V_{ba}, the multimeter's red lead would need to be placed at point b and the black lead would be placed at point a.

The other type of voltage that is frequently measured is a node voltage. A node voltage is measured between any point in the circuit and ground and may encompass many components. Node voltages are usually labeled with only one subscript. For example, the voltage between point a and ground would be labeled V_a, and V_a is the voltage across both R_1 and R_2. V_a is also the voltage across the source, V_{in}.

Depending on how complex the circuit is, a node voltage may also give you the voltage across a single component. For example, V_b is both the voltage between point b and ground, as well as the voltage across R_2.

Current:

We know that components that are in series with each other carry the same current. Therefore, when measuring current, you are essentially putting the ammeter in series with one or more components.

To correctly measure current, you need to break the circuit and insert the ammeter's two leads across the break. For example, in Figure A.1, to measure the current coming out of the source (which is also the current through R_1), you should break the circuit between the positive side of the source and R_1, say at points a and d, as seen in Figure A.2. Then, you would place the meter's red lead at point a and the black lead at point d. This is consistent with the direction of the current I_1's arrow. If the arrow for I_1 was pointing from right-to-left, you would place the meter's red lead at point d and the black lead at point a.

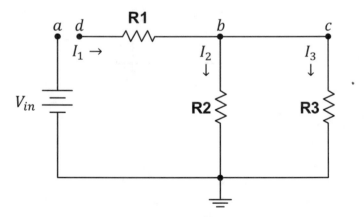

Figure A.2: A simple circuit broken to measure I_1.

APPENDIX B

Multisim Instructions

Purpose: These instructions document how to create and simulate an electronic schematic using Multisim.

Adding components: Multisim has a huge number of electronic components available to use, and adding them to a drawing is straightforward.

- To add and move a component:
  - Go to menu: Place: Component.
  - In the window that appears, chose the "Group," "Family," and "Component" from the drop-down menus. For example, if you choose Group: Basic, Family: Resistor, and Component: 1k, you are choosing a 1 kΩ resistor.
  - After the component desired is highlighted either select "OK" or double click on the item in the "Component" you desire.
  - Left-click anywhere in the drawing to place the chosen component.
  - Once placed, you can move the component by dragging with the mouse or using arrow keys.
- If you can't find the component you are looking for, you can search for it.
  - Go to menu: Place: Component.
  - In the window that appears, click on the "Search" button.
  - Enter the component's name (e.g., "inductor") in the field labeled "Function."

Editing components: Once you have placed a component in your drawing, you can re-position it, rotate or flip it, and change its value.

- To change the value of a component:
  - Double-click on the component.
  - Click on the "Value" tab and change the value (e.g., resistance, capacitance, inductance, etc.).
- To change what information is shown with a component or to give a component a label:
  - Double-click on the component.

- Click on the "Label" tab and change the "RefDes," or reference designation. This is the default label that is displayed when you place a component. You can also add or change a label, giving the component another name to be shown.
- Click on the "Display" tab and choose what you want to show up on the drawing. For example, you can choose to display only the component's value and hide the label and RefDes.
- To rotate or flip a component:
  - Right-click on the component and choose an action from the menu (e.g., "Flip vertically").

Adding wires to connect components: Once your components are placed on your schematic, you can connect them with wires. Multiple wires that cross must be joined with a **junction** in order to be connected.

- To add wires:
  - Hover the mouse over any component's terminal. When the cursor changes to a black dot, left-click.
  - Move the mouse along the path where you want to place a wire. You do not need to hold the mouse button down. As long as your cursor appears to be a black dot, you are placing a wire.
  - While you are placing the wire, you can control the number and location of the wire's bends by left-clicking where you want the bends to be.
  - When you reach another component's terminal or a junction, a red dot will appear. Left-click to make a connection. Your cursor will change back to an arrow.
- To add a junction:
  - Go to menu: Place: Junction. Click on the schematic where you want the junction to be.

Simulating circuits: You can "run," or simulate, a circuit in Multisim. Note: you should have ground connected to your circuit when you run a simulation; otherwise, your node voltages may not be as expected.

- To simulate a circuit:
  - Choose the green "play" button displayed above the drawing. The circuit will run until you choose the red "stop" button.
  - If you don't see the "play" button, you can go to menu: Simulate: Run to begin the simulation.

Adding Instruments: In order to see what is happening in your circuit during a simulation, you need to add measurement instruments, such as voltmeters, ammeters, oscilloscopes, and probes.

- To add a multimeter, function generator, or oscilloscope:
  - Go to Simulate: Instruments, and choose the instrument you wish to use.
    * You can add an instruments toolbar to your screen by going to menu: View: Toolbars: Instruments. A column of icons will appear to the right of your drawing.
- To add a voltmeter or ammeter:
  - Go to menu: Place: Component, and choose Group: Indicators.
- To add a measurement probe:
  - Go to menu: Place: Probe.
    * You can add a probe toolbar to your screen by going to menu: View: Toolbars: Place probe. A row of icons will appear above the drawing, to the right of the "play" button.

Adding text boxes: In order to add text to your drawing, go to menu: Place: Text.

APPENDIX C

Troubleshooting Tips

You successfully completed your pre-lab and came to your lab period excited to build and test the circuit you designed. Once breadboarded, however, your measurements don't match the values you calculated earlier. Or your measurements are all zero! Being an engineering student, you want to figure out for yourself what has gone wrong. So, before you call your TA over, troubleshoot your circuit yourself. Here are some things to check, double-check and try:

First Line of Defense: Here are a few common errors people make when using a breadboard.

- Is your power source turned on? Sometimes, you may have turned it off to work on your circuit (smart!) and then you forgot to turn it back on.
    - Verify that you have the correct source voltage, units, and waveform.
    - Make sure the source is physically connected to the circuit. It is not enough to connect power to just the breadboard or to the power rails.
    - Use your multimeter to measure the source voltage to verify the power supply is working correctly.
- Remove extraneous wires.
    - Extra wires clutter the circuit and increase the chance of errors.
    - Try to use the minimum number of wires possible.
- Check your circuit for shorts: did you accidentally wire a component to itself?
    - For each component make sure the two terminals are not connected to each other. This could happen from a misplaced wire or from seating the component in the incorrect orientation on the breadboard.
- Are you making your measurements correctly? Verify that you are measuring voltage and/or current correctly and that your multimeter is set to an appropriate range for units (e.g., mA vs. μA).

Second Line of Defense: After you've checked some obvious things, here are some more advanced things to verify.

- You may have an open circuit or a wiring problem. Make sure each component is pushed in all the way. Try removing and re-inserting each component into the breadboard.

- Measure each node voltage and compare with your calculations or Multisim values. If they do not match, verify that all your components have the correct value.

- Are your polarized components (e.g., power source or LED) oriented correctly?

Third Line of Defense: Check your op amp!

- Double-check the op amp's pin diagram, orientation, and connections. Did you put the op amp in the circuit upside down? Is the op amp pushed in all the way?

- Make sure the op amp is powered (that $V_{cc\pm}$ are connected).

- Are your $V_{cc\pm}$ values high enough for the output you expect?

- Measure the voltage across the inputs to the op amp. If the measurement is not zero make sure the feedback is connected properly. Try replacing the op amp.

Final Line of Defense: If all else fails, here are a few more things to try.

- Maybe your measurements don't match your calculations because your calculations are wrong. Double-check your theoretical calculations and consider verifying them with Multisim.

- Sometimes components that are damaged get returned to drawers instead of reported or thrown out. Try replacing all of your components with new ones.

- Sometimes wires have breaks in the middle. Inspect and/or replace all of your wires.

- Breadboards that get a lot of use sometimes develop "dead zones," where holes that are supposed to be connected are not connected any longer. Try moving your circuit to a different part of the breadboard or to a new breadboard.

APPENDIX D

Standard Resistor and Capacitor Values

Table D.1: Standard resistor values (± 5%) in Ohms (Ω)

10	100	1 k	10 k	100 k
11	110	1.1 k	11 k	110 k
12	120	1.2 k	12 k	120 k
13	130	1.3 k	13 k	130 k
15	150	1.5 k	15 k	150 k
16	160	1.6 k	16 k	160 k
18	180	1.8 k	18 k	180 k
20	200	2.0 k	20 k	200 k
22	220	2.2 k	22 k	220 k
24	240	2.4 k	24 k	240 k
27	270	2.7 k	27 k	270 k
30	300	3.0 k	30 k	300 k
33	330	3.3 k	33 k	330 k
36	360	3.6 k	36 k	360 k
39	390	3.9 k	39 k	390 k
43	430	4.3 k	43 k	430 k
47	470	4.7 k	47 k	470 k
51	510	5.1 k	51 k	510 k
56	560	5.6 k	56 k	560 k
62	620	6.2 k	62 k	620 k
68	680	6.8 k	68 k	680 k
75	750	7.5 k	75 k	750 k
82	820	8.2 k	82 k	820 k
91	910	9.1 k	91 k	910 k

Table D.2: Standard capacitor values in Farads (F)

1.0 p	10 p	100 p	1.0 n	10 n	100 n	1.0 μ	10 μ	100 μ
1.1 p	11 p	110 p	1.1 n	11 n	110 n			
1.2 p	12 p	120 p	1.2 n	12 n	120 n			
1.3 p	13 p	130 p	1.3 n	13 n	130 n			
1.5 p	15 p	150 p	1.5 n	15 n	150 n	1.5 μ	15 μ	150 μ
1.6 p	16 p	160 p	1.6 n	16 n	160 n			
1.8 p	18 p	180 p	1.8 n	18 n	180 n			
2.0 p	20 p	200 p	2.0 n	20 n	200 n			
2.2 p	22 p	220 p	2.2 n	22 n	220 n	2.2 μ	22 μ	220 μ
2.4 p	24 p	240 p	2.4 n	24 n	240 n			
2.7 p	27 p	270 p	2.7 n	27 n	270 n			
3.0 p	30 p	300 p	3.0 n	30 n	300 n			
3.3 p	33 p	330 p	3.3 n	33 n	330 n	3.3 μ	33 μ	330 μ
3.6 p	36 p	360 p	3.6 n	36 n	360 n			
3.9 p	39 p	390 p	3.9 n	39 n	390 n			
4.3 p	43 p	430 p	4.3 n	43 n	430 n			
4.7 p	47 p	470 p	4.7 n	47 n	470 n	4.7 μ	47 μ	470 μ
5.1 p	51 p	510 p	5.1 n	51 n	510 n			
5.6 p	56 p	560 p	5.6 n	56 n	560 n			
6.2 p	62 p	620 p	6.2 n	62 n	620 n			
6.8 p	68 p	680 p	6.8 n	68 n	680 n	6.8 μ	68 μ	680 μ
7.5 p	75 p	750 p	7.5 n	75 n	750 n			
8.2 p	82 p	820 p	8.2 n	82 n	820 n			
9.1 p	91 p	910 p	9.1 n	91 n	910 n			

APPENDIX E

Potentiometers

A *potentiometer* is a type of variable resistor that has three terminals and an adjustable dial. A potentiometer can be used by itself as a voltage divider by connecting one of the outside terminals to $V+$ and the other outside terminal to ground. The middle terminal then becomes a node that outputs a voltage between zero and $V+$ as you turn the dial. This configuration is shown in Figure E.1.

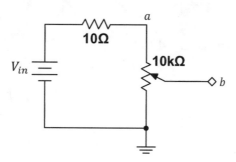

Figure E.1: Potentiometer as voltage divider.

In Figure E.1, one outside terminal of the potentiometer is connected to the 10 Ω resistor (node a), the other outside terminal is connected to ground, and the middle terminal is labeled node b. As you turn the potentiometer's dial, V_{ab} and V_b will vary: as V_{ab} increases, V_b will decrease so that V_a remains constant. So the potentiometer is essentially functioning like two, complementary variable resistors.

You can also use a potentiometer as a simple variable resistor by using only the middle terminal and one of the outside terminals (leaving the other outside terminal unconnected). This configuration is shown in Figure E.2.

In Figure E.2, one outside terminal of the potentiometer is connected to the 10 Ω resistor (node c), the other outside terminal is unconnected, and the middle terminal is connected to ground. In this configuration, when you turn the dial, the resistance between the two terminals in use varies between zero and 10 kΩ, the maximum resistance of the device. In other words, V_c will vary between zero volts (when the resistance is at its minimum, 0 Ω) and some maximum voltage (when the resistance is at its maximum value, 10 kΩ). An equivalent circuit is shown in Figure E.3.

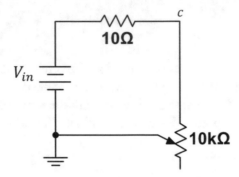

Figure E.2: Potentiometer as variable resistor.

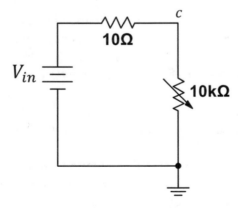

Figure E.3: Equivalent circuit using a variable resistor instead of a potentiometer.

APPENDIX F

Thermistor Data Sheet

The following data sheet is for Dale/Vishay NTC Thermistor #01C1002KP. Additional information can be found at: `https://www.vishay.com/thermistors/ntc-curve-list/`

SAP P/N: 01C1002K

description of the product
LEADED EPOXY COATED AWG 28 SOLID COPPER WIRES

product electrical characteristics characteristics

R(25°C or 77°F)	Tref	Tolerance at Tref	curve		Tmin	Tmax	B0/100 (K)	3947.00
ohms	°F	%			°F	°F	B25/75(K)	3964.00
10000.00	77	10		1	-58.00	257.00	B25/85(K)	3974.00

Steinhart & hart Coefficents for R/T computation

	A	B	C	temperature scale	°F
				temperature step	1.00
coef.valid between	1.1249E-03	2.3482E-04	8.4976E-08	start temperature	-58
0 and 100 °C				end temperature	257

R in ohms $\quad$ T in K	$\frac{1}{T} = A + B\ln(R) + C\ln^3(R)$	$R = \exp[(x - y/2)^{1/3} - (x + y/2)^{1/3}]$	$y = \frac{\left(A - \frac{1}{T}\right)}{C}$ $\quad$ $x = \sqrt{\left(\frac{B}{3C}\right)^3 + \frac{y^2}{4}}$

Electrical Resistance in function of the temperature

Temperature °F	Resistance value (ohms)	R/R25	delta R/R (%)	alpha (%/K)	dT (K)	Rmin (ohms)	Rmax (ohms)
-58.0	672600.0	67.260	13.90	-7.15	1.94	579108.6	766091.4
-57.0	648999.0	64.900	13.87	-7.12	1.95	558982.8	739015.1
-56.0	621935.9	62.194	13.83	-7.10	1.95	535897.3	707974.5
-55.0	600215.1	60.022	13.80	-7.07	1.95	517361.4	683068.8
-54.0	575360.6	57.536	13.77	-7.04	1.96	496145.0	654576.2
-53.0	555482.1	55.548	13.74	-7.01	1.96	479170.0	631794.2
-52.0	532634.4	53.263	13.70	-6.98	1.96	459652.8	605616.0
-51.0	514405.4	51.441	13.67	-6.96	1.97	444075.9	584734.9
-50.0	493389.1	49.339	13.64	-6.93	1.97	426110.6	560667.7
-49.0	473370.0	47.337	13.60	-6.90	1.97	408991.7	537748.3
-48.0	457316.9	45.732	13.57	-6.87	1.97	395259.0	519374.8
-47.0	438884.6	43.888	13.53	-6.85	1.98	379486.0	498283.3
-46.0	424120.4	42.412	13.50	-6.82	1.98	366847.2	481393.6
-45.0	407159.4	40.716	13.47	-6.79	1.98	352323.2	461995.6
-44.0	393572.5	39.357	13.44	-6.77	1.99	340684.2	446460.7
-43.0	377937.7	37.794	13.40	-6.74	1.99	327286.5	428588.9
-42.0	365440.1	36.544	13.37	-6.71	1.99	316573.4	414306.7
-41.0	351019.2	35.102	13.34	-6.69	1.99	304207.3	397831.1
-40.0	337270.0	33.727	13.30	-6.66	2.00	292413.1	382126.9
-39.0	326219.8	32.622	13.27	-6.64	2.00	282946.7	369492.8
-38.0	313514.0	31.351	13.22	-6.61	2.00	272058.1	354970.0
-37.0	303325.8	30.333	13.19	-6.58	2.00	263323.2	343328.4
-36.0	291605.4	29.161	13.15	-6.56	2.00	253271.0	329939.9
-35.0	282210.1	28.221	13.11	-6.53	2.01	245209.5	319210.6
-34.0	271377.3	27.138	13.07	-6.51	2.01	235911.0	306843.6
-33.0	262705.0	26.271	13.03	-6.48	2.01	228464.0	296946.0
-32.0	252687.2	25.269	12.99	-6.46	2.01	219858.1	285516.3
-31.0	243120.0	24.312	12.95	-6.43	2.01	211636.0	274604.0
-30.0	235424.9	23.542	12.92	-6.41	2.02	205019.8	265830.1
-29.0	226564.3	22.656	12.87	-6.39	2.02	197398.7	255729.9
-28.0	219442.9	21.944	12.84	-6.36	2.02	191270.8	247614.9
-27.0	211246.5	21.125	12.80	-6.34	2.02	184215.4	238277.6
-26.0	204667.1	20.467	12.76	-6.31	2.02	178549.5	230784.6
-25.0	197072.3	19.707	12.72	-6.29	2.02	172006.7	222137.9
-24.0	190981.3	19.098	12.68	-6.26	2.02	166757.2	215205.4
-23.0	183940.5	18.394	12.64	-6.24	2.03	160686.7	207194.2
-22.0	177210.0	17.721	12.60	-6.22	2.03	154881.5	199538.5
-21.0	171786.0	17.179	12.58	-6.19	2.03	150183.9	193388.1
-20.0	165532.4	16.553	12.55	-6.17	2.03	144766.4	186298.5
-19.0	160504.7	16.050	12.52	-6.15	2.04	140409.5	180599.9
-18.0	154707.1	15.471	12.49	-6.12	2.04	135384.1	174030.0
-17.0	150042.8	15.004	12.47	-6.10	2.04	131340.0	168745.7
-16.0	144658.0	14.466	12.44	-6.08	2.05	126669.8	162646.2
-15.0	140344.9	14.034	12.41	-6.05	2.05	122928.1	157761.7
-14.0	135340.5	13.534	12.38	-6.03	2.05	118585.4	152095.7
-13.0	130540.0	13.054	12.35	-6.01	2.05	114418.3	146661.7
-12.0	126676.7	12.668	12.33	-5.99	2.06	111063.8	142289.6
-11.0	122216.6	12.222	12.30	-5.97	2.06	107190.1	137243.2
-10.0	118627.5	11.863	12.27	-5.94	2.06	104071.9	133183.1

Temperature °F	Resistance value (ohms)	R/R25	delta R/R (%)	alpha (%/K)	dT (K)	Rmin (ohms)	Rmax (ohms)
-9.0	114481.7	11.448	12.24	-5.92	2.07	100469.1	128494.2
-8.0	111143.9	11.114	12.22	-5.90	2.07	97567.7	124720.2
-7.0	107284.4	10.728	12.19	-5.88	2.07	94211.8	120357.1
-6.0	104190.0	10.419	12.16	-5.86	2.08	91520.5	116859.5
-5.0	100595.1	10.060	12.13	-5.83	2.08	88392.9	112797.2
-4.0	97140.0	9.714	12.10	-5.81	2.08	85386.1	108893.9
-3.0	94357.3	9.436	12.07	-5.79	2.08	82968.4	105746.2
-2.0	91141.1	9.114	12.03	-5.77	2.09	80173.2	102109.0
-1.0	88556.6	8.856	12.00	-5.75	2.09	77926.3	99186.9
0.0	85561.0	8.556	11.97	-5.73	2.09	75321.1	95801.0
1.0	83147.4	8.315	11.94	-5.71	2.09	73221.3	93073.5
2.0	80352.5	8.035	11.90	-5.69	2.09	70789.0	89916.1
3.0	78097.8	7.810	11.87	-5.67	2.10	68826.0	87369.5
4.0	75489.2	7.549	11.84	-5.64	2.10	66554.3	84424.1
5.0	72990.0	7.299	11.80	-5.62	2.10	64377.2	81602.8
6.0	70966.4	7.097	11.77	-5.60	2.10	62613.6	79319.1
7.0	68624.7	6.862	11.73	-5.58	2.10	60572.3	76677.1
8.0	66734.0	6.673	11.70	-5.56	2.10	58923.5	74544.6
9.0	64548.1	6.455	11.67	-5.54	2.10	57016.7	72079.6
10.0	62789.0	6.279	11.64	-5.52	2.11	55481.6	70096.4
11.0	60745.2	6.075	11.60	-5.50	2.11	53697.5	67792.8
12.0	59093.0	5.909	11.57	-5.48	2.11	52254.8	65931.3
13.0	57181.5	5.718	11.54	-5.46	2.11	50585.0	63778.0
14.0	55350.0	5.535	11.50	-5.44	2.11	48984.8	61715.3
15.0	53861.0	5.386	11.48	-5.42	2.12	47677.7	60044.2
16.0	52136.7	5.214	11.46	-5.41	2.12	46163.9	58109.4
17.0	50748.4	5.075	11.44	-5.39	2.12	44944.8	56552.0
18.0	49138.1	4.914	11.41	-5.37	2.13	43530.4	54745.7
19.0	47839.4	4.784	11.39	-5.35	2.13	42389.5	53289.2
20.0	46330.8	4.633	11.37	-5.33	2.13	41063.9	51597.7
21.0	45115.5	4.512	11.35	-5.31	2.14	39995.8	50235.2
22.0	43701.6	4.370	11.32	-5.29	2.14	38752.8	48650.4
23.0	42340.0	4.234	11.30	-5.27	2.14	37555.6	47124.4
24.0	41237.7	4.124	11.28	-5.25	2.15	36586.1	45889.4
25.0	39959.7	3.996	11.26	-5.24	2.15	35461.8	44457.6
26.0	38931.6	3.893	11.24	-5.22	2.15	34557.3	43306.0
27.0	37734.4	3.773	11.21	-5.20	2.16	33503.7	41965.2
28.0	36765.5	3.677	11.19	-5.18	2.16	32650.7	40880.3
29.0	35641.9	3.564	11.17	-5.16	2.16	31661.4	39622.3
30.0	34734.8	3.473	11.15	-5.14	2.17	30862.5	38607.0
31.0	33679.7	3.368	11.12	-5.13	2.17	29933.1	37426.2
32.0	32660.0	3.266	11.10	-5.11	2.17	29034.7	36285.3
33.0	31837.5	3.184	11.09	-5.09	2.18	28308.3	35366.7
34.0	30882.5	3.088	11.07	-5.07	2.18	27464.8	34300.3
35.0	30111.0	3.011	11.05	-5.06	2.19	26783.1	33438.9
36.0	29213.3	2.921	11.03	-5.04	2.19	25989.9	32436.7
37.0	28484.7	2.848	11.02	-5.02	2.20	25346.0	31623.4
38.0	27640.7	2.764	11.00	-5.00	2.20	24599.9	30681.4
39.0	26956.9	2.696	10.99	-4.99	2.20	23995.4	29918.4
40.0	26162.9	2.616	10.97	-4.97	2.21	23293.4	29032.5
41.0	25400.0	2.540	10.95	-4.95	2.21	22618.7	28181.3
42.0	24780.9	2.478	10.94	-4.93	2.22	22071.1	27490.7
43.0	24061.1	2.406	10.92	-4.92	2.22	21434.4	26687.9
44.0	23474.4	2.347	10.90	-4.90	2.22	20915.2	26033.5
45.0	22795.6	2.280	10.88	-4.88	2.23	20314.6	25276.7
46.0	22241.6	2.224	10.87	-4.87	2.23	19824.2	24659.0
47.0	21602.4	2.160	10.85	-4.85	2.24	19258.3	23946.4
48.0	21092.5	2.109	10.84	-4.83	2.24	18806.9	23378.1
49.0	20489.9	2.049	10.82	-4.82	2.25	18273.3	22706.5
50.0	19900.0	1.990	10.80	-4.80	2.25	17750.8	22049.2
51.0	19428.2	1.943	10.77	-4.78	2.25	17336.8	21519.7
52.0	18879.5	1.888	10.72	-4.77	2.25	16855.1	20904.0
53.0	18436.2	1.844	10.69	-4.75	2.25	16465.7	20406.6
54.0	17919.0	1.792	10.65	-4.74	2.25	16011.3	19826.6
55.0	17505.6	1.751	10.61	-4.72	2.25	15648.0	19363.1
56.0	17017.4	1.702	10.57	-4.70	2.25	15218.8	18816.0
57.0	16624.8	1.662	10.53	-4.69	2.25	14873.6	18376.1
58.0	16164.0	1.616	10.49	-4.67	2.25	14468.1	17859.9
59.0	15710.0	1.571	10.45	-4.66	2.24	14068.3	17351.7
60.0	15350.0	1.535	10.42	-4.64	2.24	13751.3	16948.7
61.0	14930.6	1.493	10.37	-4.63	2.24	13381.9	16479.4
62.0	14585.8	1.459	10.34	-4.61	2.24	13077.9	16093.7
63.0	14188.7	1.419	10.30	-4.59	2.24	12727.8	15649.5
64.0	13868.8	1.387	10.26	-4.58	2.24	12445.7	15291.9
65.0	13493.4	1.349	10.22	-4.56	2.24	12114.5	14872.3
66.0	13189.2	1.319	10.18	-4.55	2.24	11846.0	14532.4
67.0	12834.3	1.283	10.14	-4.53	2.24	11532.7	14136.0
68.0	12490.0	1.249	10.10	-4.52	2.24	11228.5	13751.5
69.0	12211.2	1.221	10.09	-4.50	2.24	10979.1	13443.3
70.0	11886.4	1.189	10.08	-4.49	2.25	10688.5	13084.3

Temperature °F	Resistance value (ohms)	R/R25	delta R/R (%)	alpha (%/K)	dT (K)	Rmin (ohms)	Rmax (ohms)
71.0	11625.7	1.163	10.07	-4.47	2.25	10455.2	12796.1
72.0	11318.3	1.132	10.06	-4.46	2.26	10180.2	12456.5
73.0	11066.3	1.107	10.05	-4.44	2.26	9954.5	12178.0
74.0	10775.4	1.078	10.03	-4.43	2.27	9694.2	11856.6
75.0	10542.5	1.054	10.02	-4.42	2.27	9485.7	11599.3
76.0	10267.1	1.027	10.01	-4.40	2.27	9239.1	11295.0
77.0	10000.0	1.000	10.00	-4.39	2.28	9000.0	11000.0
78.0	9783.1	0.978	10.02	-4.37	2.29	8802.9	10763.4
79.0	9530.3	0.953	10.04	-4.36	2.30	8573.1	10487.5
80.0	9325.2	0.933	10.06	-4.34	2.32	8386.7	10263.7
81.0	9085.8	0.909	10.09	-4.33	2.33	8169.2	10002.4
82.0	8891.1	0.889	10.11	-4.32	2.34	7992.4	9789.8
83.0	8664.1	0.866	10.13	-4.30	2.36	7786.3	9542.0
84.0	8480.3	0.848	10.15	-4.29	2.37	7619.4	9341.2
85.0	8265.0	0.827	10.18	-4.27	2.38	7424.0	9106.1
86.0	8055.0	0.806	10.20	-4.26	2.39	7233.4	8876.6
87.0	7885.8	0.789	10.23	-4.25	2.41	7079.5	8692.2
88.0	7688.3	0.769	10.26	-4.23	2.42	6899.9	8476.8
89.0	7527.6	0.753	10.28	-4.22	2.44	6753.8	8301.5
90.0	7339.9	0.734	10.31	-4.21	2.45	6583.1	8096.6
91.0	7187.6	0.719	10.34	-4.19	2.47	6444.8	7930.5
92.0	7009.3	0.701	10.37	-4.18	2.48	6282.8	7735.9
93.0	6864.8	0.686	10.39	-4.16	2.49	6151.6	7578.1
94.0	6695.5	0.670	10.42	-4.15	2.51	5997.9	7393.2
95.0	6531.0	0.653	10.45	-4.14	2.53	5848.5	7213.5
96.0	6397.3	0.640	10.48	-4.12	2.54	5727.2	7067.5
97.0	6241.2	0.624	10.51	-4.11	2.55	5585.6	6896.9
98.0	6114.2	0.611	10.53	-4.10	2.57	5470.4	6758.0
99.0	5965.9	0.597	10.56	-4.09	2.58	5335.9	6595.9
100.0	5844.8	0.584	10.59	-4.07	2.60	5226.2	6463.5
101.0	5703.9	0.570	10.62	-4.06	2.61	5098.4	6309.4
102.0	5590.0	0.559	10.64	-4.05	2.63	4995.2	6184.8
103.0	5456.0	0.546	10.67	-4.03	2.65	4873.8	6038.1
104.0	5325.0	0.533	10.70	-4.02	2.66	4755.2	5894.8
105.0	5219.2	0.522	10.72	-4.01	2.67	4659.7	5778.7
106.0	5095.6	0.510	10.74	-4.00	2.69	4548.1	5643.0
107.0	4994.8	0.499	10.76	-3.98	2.70	4457.1	5532.4
108.0	4877.0	0.488	10.79	-3.97	2.72	4350.9	5403.2
109.0	4781.3	0.478	10.81	-3.96	2.73	4264.6	5298.1
110.0	4669.2	0.467	10.83	-3.95	2.75	4163.5	5175.0
111.0	4578.8	0.458	10.85	-3.93	2.76	4081.9	5075.7
112.0	4472.1	0.447	10.88	-3.92	2.77	3985.7	4958.5
113.0	4368.0	0.437	10.90	-3.91	2.79	3891.9	4844.1
114.0	4283.5	0.428	10.92	-3.90	2.80	3815.7	4751.3
115.0	4184.7	0.418	10.94	-3.88	2.82	3726.7	4642.7
116.0	4103.9	0.410	10.96	-3.87	2.83	3654.0	4553.9
117.0	4009.8	0.401	10.99	-3.86	2.85	3569.2	4450.4
118.0	3933.1	0.393	11.01	-3.85	2.86	3500.1	4366.0
119.0	3843.4	0.384	11.03	-3.84	2.88	3419.4	4267.4
120.0	3770.7	0.377	11.05	-3.82	2.89	3354.0	4187.4
121.0	3685.2	0.369	11.08	-3.81	2.90	3277.0	4093.4
122.0	3602.0	0.360	11.10	-3.80	2.92	3202.2	4001.8
123.0	3534.2	0.353	11.11	-3.79	2.93	3141.6	3926.9
124.0	3454.9	0.345	11.12	-3.78	2.94	3070.6	3839.2
125.0	3390.6	0.339	11.13	-3.77	2.96	3013.1	3768.0
126.0	3315.0	0.331	11.14	-3.75	2.97	2945.5	3684.4
127.0	3253.3	0.325	11.15	-3.74	2.98	2890.4	3616.1
128.0	3181.1	0.318	11.17	-3.73	2.99	2825.9	3536.3
129.0	3122.1	0.312	11.18	-3.72	3.00	2773.2	3471.0
130.0	3053.2	0.305	11.19	-3.71	3.02	2711.6	3394.8
131.0	2987.0	0.299	11.20	-3.70	3.03	2652.5	3321.5
132.0	2931.9	0.293	11.21	-3.69	3.04	2603.2	3260.5
133.0	2867.4	0.287	11.22	-3.68	3.05	2545.6	3189.2
134.0	2815.9	0.282	11.23	-3.66	3.07	2499.6	3132.2
135.0	2754.8	0.275	11.24	-3.65	3.08	2445.0	3064.5
136.0	2704.3	0.270	11.25	-3.64	3.09	2400.0	3008.7
137.0	2645.9	0.265	11.27	-3.63	3.10	2347.9	2944.0
138.0	2598.7	0.260	11.28	-3.62	3.12	2305.7	2891.7
139.0	2542.9	0.254	11.29	-3.61	3.13	2255.9	2830.0
140.0	2488.0	0.249	11.30	-3.60	3.14	2206.9	2769.1
141.0	2444.0	0.244	11.33	-3.59	3.16	2167.2	2720.8
142.0	2392.4	0.239	11.36	-3.58	3.17	2120.8	2664.1
143.0	2350.2	0.235	11.38	-3.57	3.19	2082.8	2617.7
144.0	2300.5	0.230	11.41	-3.55	3.21	2038.1	2563.0
145.0	2259.9	0.226	11.44	-3.54	3.23	2001.4	2518.3
146.0	2212.4	0.221	11.47	-3.53	3.24	1958.7	2466.0
147.0	2173.2	0.217	11.49	-3.52	3.26	1923.5	2422.9
148.0	2127.8	0.213	11.52	-3.51	3.28	1882.7	2372.9
149.0	2084.0	0.208	11.55	-3.50	3.30	1843.3	2324.7
150.0	2047.6	0.205	11.58	-3.49	3.31	1810.6	2284.6

Temperature °F	Resistance value (ohms)	R/R25	delta R/R (%)	alpha (%/K)	dT (K)	Rmin (ohms)	Rmax (ohms)
151.0	2005.0	0.200	11.61	-3.48	3.33	1772.3	2237.7
152.0	1971.1	0.197	11.63	-3.47	3.35	1741.9	2200.4
153.0	1930.6	0.193	11.66	-3.46	3.37	1705.5	2155.7
154.0	1897.5	0.190	11.69	-3.45	3.39	1675.8	2119.2
155.0	1858.7	0.186	11.72	-3.44	3.41	1640.9	2076.4
156.0	1826.5	0.183	11.74	-3.43	3.42	1612.0	2040.9
157.0	1789.3	0.179	11.77	-3.42	3.44	1578.7	1999.9
158.0	1753.0	0.175	11.80	-3.41	3.46	1546.1	1959.9
159.0	1723.7	0.172	11.82	-3.40	3.48	1520.0	1927.3
160.0	1689.3	0.169	11.83	-3.39	3.49	1489.4	1889.1
161.0	1660.3	0.166	11.85	-3.38	3.51	1463.5	1857.0
162.0	1627.0	0.163	11.87	-3.37	3.52	1433.9	1820.0
163.0	1600.0	0.160	11.88	-3.36	3.54	1409.9	1790.1
164.0	1568.1	0.157	11.90	-3.35	3.55	1381.5	1754.7
165.0	1542.3	0.154	11.91	-3.34	3.57	1358.5	1726.0
166.0	1511.7	0.151	11.93	-3.33	3.58	1331.3	1692.0
167.0	1482.0	0.148	11.95	-3.32	3.60	1304.9	1659.1
168.0	1457.2	0.146	11.97	-3.31	3.61	1282.9	1631.6
169.0	1428.3	0.143	11.98	-3.30	3.63	1257.1	1599.4
170.0	1405.4	0.141	12.00	-3.29	3.64	1236.7	1574.0
171.0	1377.9	0.138	12.02	-3.28	3.66	1212.3	1543.5
172.0	1355.2	0.136	12.03	-3.27	3.68	1192.2	1518.3
173.0	1328.9	0.133	12.05	-3.26	3.69	1168.8	1489.0
174.0	1307.5	0.131	12.06	-3.25	3.71	1149.7	1465.2
175.0	1282.2	0.128	12.08	-3.24	3.72	1127.3	1437.1
176.0	1257.0	0.126	12.10	-3.24	3.74	1104.9	1409.1
177.0	1237.3	0.124	12.11	-3.23	3.75	1087.5	1387.1
178.0	1214.1	0.121	12.12	-3.22	3.77	1066.9	1361.3
179.0	1194.2	0.119	12.13	-3.21	3.78	1049.3	1339.1
180.0	1171.5	0.117	12.14	-3.20	3.80	1029.2	1313.7
181.0	1153.0	0.115	12.15	-3.19	3.81	1012.8	1293.1
182.0	1131.1	0.113	12.17	-3.18	3.83	993.5	1268.7
183.0	1113.0	0.111	12.18	-3.17	3.84	977.5	1248.6
184.0	1092.1	0.109	12.19	-3.16	3.85	959.0	1225.2
185.0	1072.0	0.107	12.20	-3.15	3.87	941.2	1202.8
186.0	1055.3	0.106	12.21	-3.14	3.88	926.5	1184.2
187.0	1035.7	0.104	12.22	-3.14	3.90	909.2	1162.3
188.0	1019.6	0.102	12.23	-3.13	3.91	894.9	1144.4
189.0	1000.7	0.100	12.24	-3.12	3.93	878.7	1123.3
190.0	985.0	0.099	12.25	-3.11	3.94	864.3	1105.7
191.0	966.8	0.097	12.27	-3.10	3.96	848.3	1085.4
192.0	952.0	0.095	12.28	-3.09	3.97	835.1	1068.8
193.0	934.5	0.093	12.29	-3.08	3.99	819.6	1049.3
194.0	917.4	0.092	12.30	-3.07	4.00	804.6	1030.2
195.0	903.4	0.090	12.32	-3.07	4.02	792.2	1014.7
196.0	887.0	0.089	12.33	-3.06	4.03	777.6	996.4
197.0	873.6	0.087	12.35	-3.05	4.05	765.7	981.4
198.0	857.8	0.086	12.37	-3.04	4.07	751.7	963.8
199.0	844.8	0.084	12.38	-3.03	4.08	740.2	949.4
200.0	829.6	0.083	12.40	-3.02	4.10	726.8	932.5
201.0	817.2	0.082	12.41	-3.01	4.12	715.8	918.7
202.0	802.6	0.080	12.43	-3.01	4.14	702.8	902.3
203.0	788.2	0.079	12.45	-3.00	4.15	690.1	886.3
204.0	776.5	0.078	12.47	-2.99	4.17	679.7	873.3
205.0	762.7	0.076	12.48	-2.98	4.19	667.5	857.9
206.0	751.5	0.075	12.50	-2.97	4.20	657.5	845.4
207.0	738.2	0.074	12.52	-2.96	4.22	645.8	830.6
208.0	727.3	0.073	12.53	-2.96	4.24	636.2	818.5
209.0	714.5	0.071	12.55	-2.95	4.26	624.9	804.2
210.0	704.1	0.070	12.56	-2.94	4.27	615.7	792.6
211.0	691.8	0.069	12.58	-2.93	4.29	604.8	778.9
212.0	679.8	0.068	12.60	-2.92	4.31	594.1	765.5
213.0	669.9	0.067	12.61	-2.92	4.32	585.5	754.4
214.0	658.3	0.066	12.61	-2.91	4.34	575.3	741.3
215.0	648.8	0.065	12.62	-2.90	4.35	566.9	730.6
216.0	637.6	0.064	12.62	-2.89	4.37	557.1	718.1
217.0	628.5	0.063	12.63	-2.88	4.38	549.1	707.9
218.0	617.7	0.062	12.63	-2.88	4.39	539.7	695.8
219.0	608.9	0.061	12.64	-2.87	4.41	531.9	685.8
220.0	598.5	0.060	12.64	-2.86	4.42	522.8	674.2
221.0	588.4	0.059	12.65	-2.85	4.44	514.0	662.8
222.0	580.1	0.058	12.66	-2.84	4.45	506.7	653.5
223.0	570.3	0.057	12.66	-2.84	4.46	498.1	642.5
224.0	562.2	0.056	12.67	-2.83	4.48	491.0	633.4
225.0	552.8	0.055	12.67	-2.82	4.49	482.7	622.8
226.0	545.0	0.054	12.68	-2.81	4.51	475.9	614.1
227.0	535.9	0.054	12.68	-2.81	4.52	467.9	603.8
228.0	528.4	0.053	12.69	-2.80	4.53	461.4	595.5
229.0	519.7	0.052	12.69	-2.79	4.55	453.7	585.6
230.0	511.0	0.051	12.70	-2.78	4.56	446.1	575.9

Temperature °F	Resistance value (ohms)	R/R25	delta R/R (%)	alpha (%/K)	dT (K)	Rmin (ohms)	Rmax (ohms)
231.0	503.9	0.050	12.73	-2.78	4.58	439.8	568.1
232.0	495.6	0.050	12.76	-2.77	4.61	432.4	558.8
233.0	488.9	0.049	12.78	-2.76	4.63	426.4	551.3
234.0	480.8	0.048	12.81	-2.75	4.65	419.2	542.4
235.0	474.3	0.047	12.84	-2.75	4.67	413.4	535.2
236.0	466.5	0.047	12.87	-2.74	4.70	406.5	526.6
237.0	460.2	0.046	12.89	-2.73	4.72	400.9	519.5
238.0	452.7	0.045	12.92	-2.72	4.74	387.7	511.2
239.0	445.4	0.045	12.95	-2.72	4.77	387.7	503.1
240.0	439.4	0.044	12.98	-2.71	4.79	382.4	496.4
241.0	432.3	0.043	13.01	-2.70	4.81	376.1	488.6
242.0	426.5	0.043	13.03	-2.70	4.83	370.9	482.0
243.0	419.6	0.042	13.06	-2.69	4.86	364.8	474.4
244.0	414.1	0.041	13.09	-2.68	4.88	359.9	468.3
245.0	407.5	0.041	13.12	-2.67	4.91	354.1	461.0
246.0	402.0	0.040	13.14	-2.67	4.93	349.2	454.9
247.0	395.7	0.040	13.17	-2.66	4.95	343.6	447.8
248.0	389.4	0.039	13.20	-2.65	4.98	338.0	440.8
249.0	384.3	0.038	13.21	-2.65	4.99	333.5	435.1
250.0	378.3	0.038	13.22	-2.64	5.01	328.3	428.3
251.0	373.3	0.037	13.23	-2.63	5.03	323.9	422.7
252.0	367.5	0.037	13.24	-2.62	5.05	318.8	416.1
253.0	362.6	0.036	13.25	-2.62	5.06	314.6	410.7
254.0	357.0	0.036	13.27	-2.61	5.08	309.6	404.3
255.0	352.4	0.035	13.28	-2.60	5.10	305.6	399.2
256.0	347.0	0.035	13.29	-2.60	5.12	300.9	393.1
257.0	341.6	0.034	13.30	-2.59	5.13	296.2	387.0

APPENDIX G

Thevenin and Norton Analysis

To find a Thevenin or Norton equivalent circuit, there are two, and sometimes three, steps that must be followed. In the example circuit shown in Figure G.1, the 6 Ω resistor is the load, and we want to find an equivalent circuit.

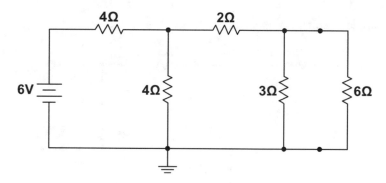

Figure G.1: Example circuit with 6 Ω resistor as load.

Step 1: Find the open circuit voltage, V_{oc}

A logical first step in these calculations is to find the Thevenin voltage, V_{th}, also known as the open circuit voltage, V_{oc}. To do so, follow these steps.

- Open the load resistor. (Remove the load resistor from the circuit.) See Figure G.2.

- Find the voltage, V_{oc}, across the open. Using node analysis is often useful.

For our example $V_{oc} = V_{th} = 9/7$ V.

Step 2: Find the Thevenin resistance, R_{th}

If there are no dependent sources in the circuit, you can calculate R_{th}. To do so, follow these steps.

- Open the load resistor. (Remove the load resistor, as you did when calculating V_{oc}.)

- "Turn off," or set to zero, all independent sources in the circuit. This means replacing any independent voltage sources with a short and replacing any independent current sources with an open circuit. See Figure G.3.

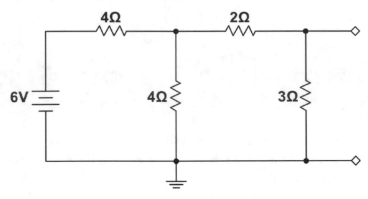

Figure G.2: Example circuit with opened load.

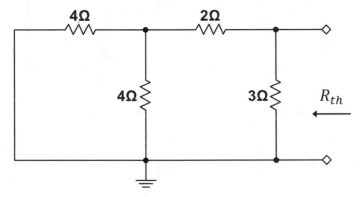

Figure G.3: Example circuit with opened load and shorted voltage source.

- Combine all resistors to find a single equivalent resistance as seen from the open terminals. It is usually wise to begin combining resistors opposite to the end where the open is located.

For our example, $R_{th} = 12/7 \ \Omega$.

Step 3: Find the short circuit current, i_{sc}

If there are dependent sources in the circuit, you cannot zero them out or remove them, and, therefore, you cannot directly find the Thevenin resistance. Instead, find i_{sc}. To do so, follow these steps.

- Short the load resistor. (Replace the load resistor with a wire.) See Figure G.4.
- Find the current, i_{sc}, through the short. Using mesh/loop analysis is often useful. For our example, $i_{sc} = 3/4$ A.

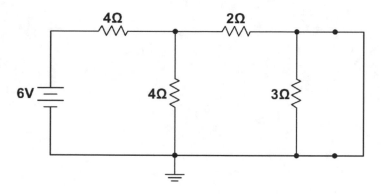

Figure G.4: Example circuit with shorted load.

Step 4: Use Ohm's law to calculate the missing quantity

Using the fact that $V_{oc} = i_{sc} R_{th}$, find any value that was not solved for above. If you solved for V_{oc} and R_{th}, you can draw the Thevenin equivalent circuit (shown in Figure G.5). If you solved for i_{sc} and R_{th}, you can draw the Norton equivalent circuit (shown in Figure G.6). If there were dependent sources in the original circuit and you calculated V_{oc} and i_{sc}, use Ohm's law to calculate R_{th}, and you are ready to draw the Thevenin or Norton equivalent circuit.

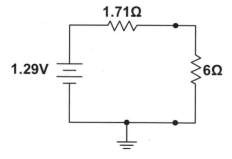

Figure G.5: Thevenin equivalent of example circuit.

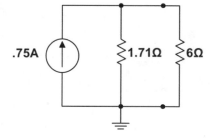

Figure G.6: Norton equivalent of example circuit.

Bibliography

[1] Unless otherwise noted, all figures, photographs, and visuals were created by the authors.

[2] Munroe, Randall. xkcd: Snakes. *xkcd*. Web. December 28, 2015. `https://xkcd.com/1604`

[3] Photo appears courtesy of Thomas R. Beard.

Authors' Biographies

TERI L. PIATT

Teri L. Piatt holds a Bachelor of Science in electrical engineering from the University of Notre Dame and a Master of Science and Ph.D. in electrical engineering from the University of Colorado, Boulder. She has worked for GE Aviation and for Jacobs Sverdrup as a research engineer in the Sensors Directorate at the Air Force Research Laboratory. She is currently a Lecturer at Wright State University.

KYLE E. LAFERTY

Kyle E. Laferty graduated *summa cum laude* from Wright State University with a Bachelor of Science in electrical engineering and a minor in mathematics and is a Ph.D. student at the University of Michigan.

Printed in the United States
by Baker & Taylor Publisher Services